Health and the Environment in the Southeastern United States

Howard Frumkin, Richard J. Jackson, and Christine M. Coussens, *Editors*

Roundtable on Environmental Health Sciences, Research, and Medicine

Division of Health Sciences Policy

INSTITUTE OF MEDICINE
OF THE NATIONAL ACADEMIES

THE NATIONAL ACADEMIES PRESS
Washington, D.C.
www.nap.edu

THE NATIONAL ACADEMIES PRESS • 500 Fifth Street, N.W. • Washington, DC 20001

NOTICE: The project that is the subject of this report was approved by the Governing Board of the National Research Council, whose members are drawn from the councils of the National Academy of Sciences, the National Academy of Engineering, and the Institute of Medicine. The members of the committee responsible for the report were chosen for their special competences and with regard for appropriate balance.

Support for this project was provided by the National Institute of Environmental Health Sciences, National Institute of Health (Contract No. 282-99-0045, TO#5); National Center for Environmental Health and Agency for Toxic Substances and Disease Registry, Centers for Disease Control and Prevention (Contract No. 200-2000-00629, TO#7); National Institute for Occupational Safety and Health, Centers for Disease Control and Prevention, (Contract No. 0000166930); National Health and Environment Effects Research Laboratory and National Center for Environmental Research, Environmental Protection Agency (Contract No. 282-99-0045, TO#5); American Chemistry Council (unnumbered grant); and Exxon-Mobil Corporation (unnumbered grant). The views presented in this report are those of the individual presenters and are not necessarily those of the funding agencies or the Institute of Medicine.

This summary is based on the proceedings of a workshop that was sponsored by the Roundtable on Environmental Health Sciences, Research, and Medicine. It is prepared in the form of a workshop summary by and in the name of the ditors, with the assistance of staff and consultants, as an individuall authored document.

Additional copies of this report are available for sale from the National Academies Press, 500 Fifth Street, NW, Lock box 285, Washington, DC, 20055; (800) 624-6242 or (202) 334-3313 (in the Washington metropolitan area); Internet, http://www.nap.edu.

For more information about the Institute of Medicine, visit the IOM home page at: **www.iom.edu.**

Copyright 2002 by the National Academy of Sciences. All rights reserved.

Printed in the United States of America.

The serpent has been a symbol of long life, healing, and knowledge among almost all cultures and religions since the beginning of recorded history. The serpent adopted as a logotype by the Institute of Medicine is a relief carving from ancient Greece, now held by the Staatliche Museen in Berlin.

*"Knowing is not enough; we must apply.
Willing is not enough; we must do."*
—Goethe

INSTITUTE OF MEDICINE
OF THE NATIONAL ACADEMIES

Shaping the Future for Health

THE NATIONAL ACADEMIES
Advisers to the Nation on Science, Engineering, and Medicine

The **National Academy of Sciences** is a private, nonprofit, self-perpetuating society of distinguished scholars engaged in scientific and engineering research, dedicated to the furtherance of science and technology and to their use for the general welfare. Upon the authority of the charter granted to it by the Congress in 1863, the Academy has a mandate that requires it to advise the federal government on scientific and technical matters. Dr. Bruce M. Alberts is president of the National Academy of Sciences.

The **National Academy of Engineering** was established in 1964, under the charter of the National Academy of Sciences, as a parallel organization of outstanding engineers. It is autonomous in its administration and in the selection of its members, sharing with the National Academy of Sciences the responsibility for advising the federal government. The National Academy of Engineering also sponsors engineering programs aimed at meeting national needs, encourages education and research, and recognizes the superior achievements of engineers. Dr. Wm. A. Wulf is president of the National Academy of Engineering.

The **Institute of Medicine** was established in 1970 by the National Academy of Sciences to secure the services of eminent members of appropriate professions in the examination of policy matters pertaining to the health of the public. The Institute acts under the responsibility given to the National Academy of Sciences by its congressional charter to be an adviser to the federal government and, upon its own initiative, to identify issues of medical care, research, and education. Dr. Harvey V. Fineberg is president of the Institute of Medicine.

The **National Research Council** was organized by the National Academy of Sciences in 1916 to associate the broad community of science and technology with the Academy's purposes of furthering knowledge and advising the federal government. Functioning in accordance with general policies determined by the Academy, the Council has become the principal operating agency of both the National Academy of Sciences and the National Academy of Engineering in providing services to the government, the public, and the scientific and engineering communities. The Council is administered jointly by both Academies and the Institute of Medicine. Dr. Bruce M. Alberts and Dr. Wm. A. Wulf are chair and vice chair, respectively, of the National Research Council.

www.national-academies.org

ROUNDTABLE ON ENVIRONMENTAL HEALTH SCIENCES, RESEARCH, AND MEDICINE
(until March 31, 2002)

Paul Grant Rogers (Chair), Partner, Hogan & Hartson, Washington, D.C.

Lynn Goldman (Vice-Chair), Professor, Johns Hopkins University Bloomberg School of Public Health, Baltimore, MD

Pauline Abernathy, Program Officer, The Pew Charitable Trusts, Philadelphia, PA

M. Brownell Anderson, Associate Vice President for Medical Education, Association of American Medical Colleges, Washington, D.C.

Roger Bulger, President and CEO, Association of Academic Health Centers, Washington, D.C.

Mark Cullen, Professor of Medicine and Public Health, Yale Occupational and Environmental Medicine Program, Yale University School of Medicine, New Haven, CT

Ruth Etzel, Editor of the American Academy of Pediatrics *Handbook of Pediatric Environmental Health*, Adjunct Professor in the Department of Environmental and Occupational Health at the George Washington University School of Public Health and Health Services, Washington, D.C.

Henry Falk, Assistant Administrator, Agency for Toxic Substance and Disease Registry, Atlanta, GA

Baruch Fischhoff, Professor of Social and Decision Sciences, Professor of Engineering & Public Policy, Department of Social and Decision Sciences, Carnegie Mellon University, Pittsburgh, PA

Howard Frumkin, Professor and Chair of the Department of Environmental and Occupational Health at Emory University's Rollins School of Public Health, Director of the Southeast Pediatric Environmental Health Specialty Unit, Atlanta, GA

Bernard D. Goldstein, Dean, University of Pittsburgh Graduate School of Public Health, Pittsburgh, PA

Robert Graham, Director, Center for Practice and Technology Assessment, Agency for Healthcare Research and Quality, Bethesda, MD

John T. Grupenhoff, President, Science and Health Communications Group, Inc., Bethesda, MD

Carol Henry, Vice President for Science and Research, American Chemistry Council

Richard J. Jackson, Director, National Center for Environmental Health, Centers for Disease Control and Prevention, Atlanta, GA

Lovell Jones, Professor, Gynecologic Oncology, University of Texas, Houston, TX

Patricia G. Kenworthy, Vice-President for Policy and Research and Senior Staff Attorney, National Environmental Trust, Washington, D.C.

Donald Mattison, Professor, Columbia University, NY
Roger McClellan, President Emeritus, Chemical Industry Institute of Toxicology, Albuquerque, NM
Sanford Miller, Senior Fellow and Adjunct Professor, Georgetown University, Washington, D.C.
Frank Mirer, Director, Health and Safety, International Union, United Auto Workers, Detroit, MI
Alan R. Nelson, Special Advisor to the CEO, American College of Physicians–American Society of Internal Medicine, Fairfax, VA
Peter Preuss, Director, National Center for Environmental Research, U.S. Environmental Protection Agency, Washington, D.C.
Lawrence Reiter, Director, National Health and Environmental Effects Research Laboratory, U.S. Environmental Protection Agency, Research Triangle Park, NC
Kathleen Rest, Acting Director, National Institute of Occupational Safety and Health, Washington, D.C.
Kenneth Olden, Director, National Institute of Environmental Health Sciences, National Institutes of Health, Research Triangle Park, N.C.
Samuel H. Wilson, Deputy Director, National Institute of Environmental Health Sciences, National Institutes of Health, Research Triangle Park, N.C.

ROUNDTABLE ON ENVIRONMENTAL HEALTH SCIENCES, RESEARCH, AND MEDICINE
(Membership April 1, 2002 to present)

Paul Grant Rogers (Chair), Partner, Hogan & Hartson, Washington, DC
Lynn Goldman (Vice-Chair), Professor, Bloomberg School of Public Health, The Johns Hopkins University, Baltimore, MD
Jacquelyne Agnew, Professor, Bloomberg School of Public Health, The Johns Hopkins University, Baltimore, MD
Jack Azar, Vice President, Environment, Health and Safety, Xerox Corporation, Webster, NY
Sophie Balk, Chairperson, Committee on Environmental Health, American Academy of Pediatrics, Bronx, NY
Roger Bulger, President and CEO, Association of Academic Health Centers, Washington, DC
Henry Falk, Assistant Administrator, Agency for Toxic Substance and Disease Registry, Centers for Disease Control and Prevention, Atlanta, GA
Baruch Fischhoff, Professor, Department of Engineering & Public Policy and the Department of Social and Decision Sciences, Carnegie Mellon University, Pittsburgh, PA

John Froines, Professor and Director, Center for Occupational and Environmental Health, Southern California Particle Center and Supersite, University of California, Los Angeles, CA

Howard Frumkin, Professor and Chair of the Department of Environmental and Occupational Health at Emory University's Rollins School of Public Health, Atlanta, GA

Michael Gallo, Professor of Environmental and Community Medicine, University of Medicine and Dentistry, New Jersey–Robert Wood Johnson Medical School, Piscataway, NJ

Bernard Goldstein, Dean, University of Pittsburgh Graduate School of Public Health, Pittsburgh, PA

Robert Graham, Director, Center for Practice and Technology Assessment, Agency for Healthcare Research and Quality, Rockville, MD

Charles Groat, Director, U.S. Geological Survey, Reston, VA

Myron Harrison, Senior Health Advisor, Exxon-Mobil, Inc., Irving, TX

Carol Henry, Vice President for Science and Research, American Chemistry Council, Arlington, VA

Richard Jackson, Director, National Center for Environmental Health, Centers for Disease Control and Prevention, Atlanta, GA

Lovell Jones, Director, Center for Research on Minority Health; Professor, Gynecologic Oncology, University of Texas, M.D. Anderson Cancer Center, Houston, TX

Alexis Karolides, Senior Research Associate, Rocky Mountain Institute, Snowmass, CO

Donald Mattison, Chairperson, Division of Epidemiology, Statistics, and Prevention Research, National Institute of Child Health and Human Development, National Institutes of Health, Bethesda, MD

Michael McGinnis, Senior Vice President and Director of the Health Group, Robert Wood Johnson Foundation, Princeton, NJ

James Melius, Director, Division of Occupational Health and Environmental Epidemiology, New York State Department of Health, New York, NY

James Merchant, Professor and Dean, College of Public Health, Iowa University, Iowa City, IA

Sanford Miller, Senior Fellow, Center for Food and Nutrition Policy, Virginia Polytechnic Institute and State University, Alexandria, VA

Alan R. Nelson, Special Advisor to the CEO, American College of Physicians–American Society of Internal Medicine, Fairfax, VA

Kenneth Olden, Director, National Institute of Environmental Health Sciences (NIEHS), National Institutes of Health, Research Triangle Park, NC

Peter Preuss, Director, National Center for Environmental Research, U.S. Environmental Protection Agency, Washington, DC

Lawrence Reiter, Director, National Health and Environmental Effects Research Laboratory, U.S. Environmental Protection Agency, Research Triangle Park, NC
Kathleen Rest, Acting Director, National Institute of Occupational Safety and Health, Centers for Disease Control and Prevention, Washington, D.C.
Samuel Wilson, Deputy Director, National Institute of Environmental Health Sciences, National Institutes of Health, Research Triangle Park, NC

IOM Health Sciences Policy Board Liaisons

Mark Cullen, Professor of medicine and public health, Yale Occupational and Environmental Medicine Program, Yale University, School of Medicine, New Haven, CT
Bernard D. Goldstein, Dean of the University of Pittsburgh Graduate School of Public Health, Pittsburgh, PA

Study Staff

Christine Coussens, Study Director
Dalia Gilbert, Research Associate
Jennifer Zavislak, Senior Project Assistant

Division Staff

Andrew Pope, Division Director
Troy Prince, Administrative Assistant
Carlos Gabriel, Financial Associate

Timothy J. Teyler, Consultant
Laurie Yelle, Consultant

REVIEWERS

This report has been reviewed in draft form by individuals chosen for their diverse perspectives and technical expertise, in accordance with procedures approved by the NRC's Report Review Committee. The purpose of this independent review is to provide candid and critical comments that will assist the institution in making its published summary as sound as possible and to ensure that the report meets institutional standards for objectivity, evidence, and responsiveness to the study charge. The review comments and draft manuscript remain confidential to protect the integrity of the deliberative process. We wish to thank the following individuals for their review of this report:

Dr. Ed Arnold, Executive Director, Physicians for Social Responsibility, Atlanta, GA

Dr. James E. Dale, Director, Environmental Health Services, Jefferson County Department of Health and Environment, CO

Dr. Camille Grayson, Director of Health Policy, Medical Association of Georgia, Atlanta, GA

Although the reviewers listed above have provided many constructive comments and suggestions, they were not asked to endorse the remarks made nor did they see the final draft of the summary before its release. The review of this report was overseen by **Melvin Worth**, Scholar-in-Residence, Institute of Medicine, who was responsible for making certain that an independent examination of this report was carried out in accordance with institutional procedures and that all review comments were carefully considered. Responsibility for the final content of this summary rests entirely with the editors and the institution.

Preface

At a workshop sponsored by the Institute of Medicine's Roundtable on Environmental Health Sciences, Research, and Medicine in June 2000, *Rebuilding the Unity of Health and the Environment: A New Vision of Environmental Health for the 21st Century*, many participants expressed the view that for a long time the world of environment, environmental regulation, environmental control, and engineering had moved in one direction, while the world of health had moved in another. From this realization arose the concept of holding a series of workshops on rebuilding the unity of health and the environment in various regions of the United States. The purpose was to bring representatives from the two worlds together to address issues of health and environment specific to each region.

The southeastern United States, which includes North Carolina, South Carolina, Georgia, Florida, Alabama, Mississippi, Tennessee, and Kentucky, was chosen to be the site of the first regional workshop. The Southeast was selected to spearhead the series of workshops because the region has a long history of confronting environmental health problems, leading environmental justice struggles, and facing new environmental challenges. The first regional workshop, *Rebuilding the Unity of Health and the Environment in the Southeastern United States*, was held in Atlanta, Georgia, on June 27, 2001.

The history of the environment and of environmental health in the Southeast is unique and very different from that of other areas of the United States. It is complex and intricately intertwined with the rise of agriculture, plantation life, industrial development, the environmental movement, the civil rights movement, and the environmental justice movement. It is also closely tied to the hot, humid climate of the region.

The natural environment of the Southeast has changed dramatically since precolonial times. Before colonization, forests of long-leaf pine and other species, pristine waterways, and a diverse topography from mountains to coastal plain dominated the landscape of the South, providing a thriving habitat for birds and a diverse range of other wildlife. The pursuits of the early colonists brought few changes to the environment, but in the eighteenth century, the untamed forests and grasslands gave way to large agricultural tracts. In the nineteenth century, agriculture was gradually overtaken by industrialization and the growth of cities. By the twentieth century, industrialization had begun to reshape the landscape of the Southeast and signaled unrelenting environmental deterioration.

Large-scale agriculture and industrial development evolved after the Civil War and into the twentieth century, further transfiguring the southern land. National corporate interests drove regional practices, from large-scale farming of single crops such as cotton to low-wage industries that relocated from the North. Much of the topsoil in the Piedmont, the hill region of the Southeast, was eroded as a result of poor farming practices. Forests were further decimated as the lumber and mill industries flourished. The marked industrial growth brought rapid population growth, which later spawned urban ghettos and noisy manufacturing towns. The post-World War II years brought the expansion of compact cities and towns into modern urban areas of economic prosperity and immense sprawl. A giant technological step for the South was the introduction of air conditioning, which led to the "Sunbelt" phenomenon, characterized by a land boom, industrial and economic development, and growth of the recreation sector.

Recent changes in the region—rapid population growth, rapid suburban development, and economic prosperity—have profoundly transformed resource use. An emerging megalopolis (dubbed "Charlantingham" by some) stretches along interstate highways from North Carolina to Alabama and includes the metropolitan areas of Charlotte, Atlanta, and Birmingham. In contrast to other major urban centers, economic and population growth is unimpeded by immediate geographical constraints, such as coastlines, water bodies, or mountains. While cities throughout the country have grown in a "sprawling" manner, the Sunbelt cities, stretching from Charlotte and Atlanta across to Phoenix, Houston, and Los Angeles, have led this trend.

Despite far-reaching changes in the environment throughout the centuries, people in the South have maintained a deep connection to the land and the waterways, which are evocative and uniquely beautiful. The southern American writer William Faulkner acknowledged this connection when he suggested that the South was "the only authentic region in the United States, because [in the South] a deep indestructible bond still exists between man and his environment" (Meriwether and Mitigate, 1988). A dominant theme throughout many of Faulkner's stories is that human life can harm the environment and that people face the choice of destroying nature or respecting it.

During the early centuries of this country's history, health in the South was deeply connected to the natural environment. Natural conditions—long summers with high heat and humidity, mild winters, and undrained ponds and swamps—enabled insects and other disease-bearing organisms to thrive. Both white and black southerners were vulnerable to epidemics and endemic diseases. Yellow fever and malaria were both serious killers. Diseases such as smallpox and tuberculosis were major problems for people of all socioeconomic levels. Pulmonary diseases such as pneumonia and pleurisy were also common. Further, the harsh working conditions on the plantations often contributed to the early deaths of the slaves. Physicians were in short supply, and medical knowledge was poor.

By the early part of the twentieth century, environmental disease had become stratified by socioeconomic level and occupation. The establishment of textile manufacturing in North Carolina, South Carolina, and Georgia brought increased urbanization and pollution. People lived in cramped conditions in compact areas with poor sanitation, and they suffered from inadequate nutrition and fatigue. Diseases such as smallpox, tuberculosis, and malaria continued to plague members of the lower socioeconomic classes, partly because of inadequate treatment. Other common diseases were uncinariasis (hookworm), caused by poor sanitation, and pellagra, caused by poor diet. Workers also suffered from occupational disorders, such as hearing loss caused by exposure to noise from heavy machinery, and byssinosis, a lung disease caused by the inhalation of cotton dust in textile mills.

The environmental movement in the United States began in the second half of the nineteenth century, as both the public and government officials awoke to the need to save the nation's wildlife heritage, restore disturbed environments, and set aside forestland and open land either for future use or for its aesthetic values. These reflections led to the organization of conservation clubs, such as the Sierra Club in New York and the National Audubon Society in Massachusetts. Much of the conservation initiative arose in the upper strata of society in the Northeast, and there was little environmental activism in the South.

By the 1960s and early 1970s, a large segment of the American public had come to realize that open spaces and wilderness areas were shrinking dramatically. The American writer Rachel Carson, in *Silent Spring,* impressed on the public that the preservation of wilderness areas and wildlife refuges would not protect the natural environment from the harmful effects of pollution. She also posited a strong link between pollution, natural resources, and human health.

Within the conservation movement in the United States, environmentalists often described environmental issues as affecting everyone equally, in an attempt to build the broadest possible constituency. Their view was that all people lived in the same biosphere, breathed the same thin layer of air, ate food grown in the same type of soil, and drew water from the same aquifers. As these issues were examined more closely, however, massive inequities in environmental exposures became evident, as did injustices in the policies used to control them. Though created equal, all Americans were not being poisoned equally.

People of color throughout the United States had long suspected that industry was targeting their neighborhoods for the most polluting businesses. By 1982, it was time for them to prove their case. The selection of a poor, predominately black community in North Carolina for a massive toxic waste dump led to public demonstrations that resulted in more than 500 arrests. Among those arrested was a prominent civil rights leader, the Reverend Joseph Lowery, one of the speakers at the Southeast regional workshop. The community's resistance to hazardous waste disposal was a new phenomenon. It brought to the forefront the issue of the environment and health and its unequal impact on the poor, and it marked the start of the environmental justice movement.

Community leaders and academic researchers in the South initiated studies demonstrating that commercial hazardous waste dump sites were disproportionately located in communities of color. Such findings shocked the members of these communities, and thousands of people of color turned into environmentalists almost overnight. Civil rights leaders and activists responded by joining grassroots environmentalist groups or forming new local environmental organizations.

Churchgoing, more prevalent in the South than in other parts of the country, provided a further arena in which environmental concerns were addressed. Church leaders often emphasized to their congregations one's natural relationship to, and partnership with, the environment. Many churchgoers became involved in environmental efforts as a reflection of their stewardship of and reverence toward the earth.

Today, issues such as resource conservation, wilderness preservation, public health reform, population control, energy conservation, antipollution regulation, and occupational health have become public health concerns in the South and throughout the country. The environmental imagination has touched nearly every institution in American society, and the word "environmental" has been attached to a range of disciplines such as law, biology, and ethics. Environmental philosophy and policy have become the concerns of millions of Americans. Yet we still face enormous challenges, and much work needs to be done, particularly in linking the environment and health.

The purpose of this regional workshop in the Southeast was to broaden the environmental health perspective from its typical focus on environmental toxicology to a view that included the impact of the natural, built, and social environments on human health. Early in the planning, Roundtable members realized that the process of engaging speakers and developing an agenda for the workshop would be nearly as instructive as the workshop itself. In their efforts to encourage a wide scope of participation, Roundtable members sought input from individuals from a broad range of diverse fields—urban planners, transportation engineers, landscape architects, developers, clergy, local elected officials, heads of industry, and others.

When approached initially, many speakers questioned whether they had anything relevant to contribute to such a workshop. As the workshop unfolded, and as participants spoke from their diverse perspectives and exchanged ideas, it became eminently clear that all were indeed "public health officers," each with vital knowledge and unique insights to offer in solving environmental health problems in the Southeast. We would like to thank this group of individuals for their immense contributions to making this meeting such a success.

This workshop summary captures the discussions that occurred during the two-day meeting. During this workshop, four main themes were explored: (1) environmental and individual health are intrinsically intertwined; (2) traditional methods of ensuring environmental health protection, such as regulations, should be balanced by more cooperative approaches to problem solving; (3) environmental health efforts should be holistic and interdisciplinary; and (4) technological advances, along with coordinated action across educational, business, social, and political spheres, offer great hope for protecting environmental health. This workshop report is an informational document that provides a summary of the regional meeting. The views expressed here do not necessarily reflect the views of the Institute of Medicine, the Roundtable, or its sponsors.

Richard J. Jackson and Howard Frumkin
Spring 2002

Contents

PREFACE xi

SUMMARY 1
    What Is Environmental Health and Where Does It Happen?, 1
    What Are Our Research Needs?, 3
    Where Do We Go From Here?, 4

1   **PERSPECTIVE ON ENVIRONMENTAL HEALTH** 7

2   **REBUILDING THE UNITY OF HEALTH AND
THE ENVIRONMENT** 10

3   **ENVIRONMENTAL HEALTH:
A FIFTY-YEAR PERSPECTIVE** 19
    The 1950s in Boston, 19
    2001 in Atlanta, 20

4   **HUMAN HEALTH AND THE
NATURAL ENVIRONMENT** 22
    Valuing the Natural Environment, 23
    Protecting the Natural Environment: Lessons from Nature, 25
    Ensuring the Health of the Natural Environment: Potential
      Strategies, 26

5   **HUMAN HEALTH AND THE BUILT ENVIRONMENT** 29
    Transportation and Healthy Environments, 30
    The Built Environment and Health Problems, 32

Environmentally Friendly Buildings, 36
Partnerships with Academia, 39
Partnerships with Industry: Creating Trust, 40
Building Healthier Cities, 41

6 HUMAN HEALTH AND THE SOCIAL ENVIRONMENT 44
Social Capital, 44
Environmental Justice, 46

REFERENCES 48

APPENDIXES

A Agenda 53

B Speakers and Panelists 57

C Meeting Participants 59

Summary*

Howard Frumkin

In planning today's regional workshop on rebuilding the unity of health and the environment in the southeastern United States, the Roundtable posed several key questions to be addressed by the participants and respondents: What is environmental health, and where does it happen? What aspects of environmental health do we have to understand better? Where do we have research needs? Where do we go from here? The information and insights offered by our speakers have provided many answers that are relevant not only to the Southeast but also to the entire country.

WHAT IS ENVIRONMENTAL HEALTH AND WHERE DOES IT HAPPEN?

A definition of environmental health begins with the definitions of "health" and "environment." The World Health Organization has defined health as more than the absence of disease or infirmity (World Health Organization, 1986); health also comprises physical, mental, and social well-being. In the workshop, many speakers and participants alluded to this definition of health. Further, they discussed health not only in conventional terms, but also in terms of livability, domestic tranquility, and social connectedness. Thus, health extends beyond biological health to encompass the condition of our society and the built communities in which we live. The environment comprises the circumstances, objects, or conditions by which we are surrounded—not only the complex of physical, chemical, and biotic factors, but also the social and cultural conditions that influence our lives and the life of our communities. Environmental health can be viewed as either the well-being of the environment or the health of individuals with respect to environmental exposures and conditions. Although some may see

---

*The summary is an edited transcript of Dr. Howard Frumkin's summations at the Atlanta meeting.

a conflict between the two views, others regard them as two aspects of the same issue. What has emerged as a unifying theme in today's workshop is the concept that healthy people exist in a healthy environment. As we maximize one, we maximize the other.

Where does environmental health take place? The concept of scales, borrowed from the field of ecology, helps us to understand that environmental health takes place simultaneously on many scales, or levels, ranging from microscopic to global.

Environmental health takes place on the molecular level. Sam Wilson explained that the combination of an individual's genetic makeup and environmental exposures determines his or her susceptibility to the ill effects of pollutants. Also, the mechanism of toxicity—the way in which toxic exposures cause harm—is largely a series of molecular and cellular events, which basic research continues to elucidate.

Environmental health occurs at the cellular level and at the organ system level. Wayne Alexander described how particulate matter in the air affects lung function, cardiovascular health, and other aspects of our health. As another example, he described how stimulating blood flow in the vessels by walking enhances cardiovascular health.

Environmental health takes place on the individual level. In recounting the high rates of automobile crashes and auto–pedestrian injuries, Ricardo Martinez reminded us of an environmental health risk faced by millions of individuals in our urban communities every day. Several participants discussed the connection between environmental conditions in our country and the rising number of individuals with asthma, cardiovascular problems, diabetes, and obesity.

Environmental health occurs on the level of family units. Dennis Creech explained the benefits of a well-sealed house in protecting occupants from diseases related to condensation.

Environmental health is evident at the community level. Dennis Creech described an office building with features that are not only healthful for workers but also protective of the environment. Michel Kilgallon explained that urban development patterns directly affect people's life-style and behaviors, such as their level of physical exercise.

Environmental health transpires at the regional level. Robert Kerr described regional programs for developing synergistic methods of waste reduction among industries. Several participants discussed regional policies that govern transportation networks, energy generation, air quality, and water quality, all of which directly affect health.

Environmental health can be appreciated at the national level. Reverend Lowery mentioned national laws mandating clean water, clean air, and waste disposal that were milestones for environmental health in our country. Other speakers discussed the regulatory agencies that work at the national level to protect the environment.

Finally, environmental health is a global phenomenon. The Reverend Lowery reminded us that our health in the largest sense emanates from, and is safeguarded by, a deep reverence for the entire creation and all who inhabit it.

From the molecular to the global, each scale is intimately connected with all of the others. Consider, for example, an individual who is involved in an automobile crash at a busy urban intersection. The crash may have occurred because of inadequate regional policies governing the design of the highway. Yet the man's life is spared because his automobile met national safety standards and he was obeying local laws mandating seat belt use. Healing of his injuries will be compromised on the molecular and cellular levels because he has diabetes. He has difficulty controlling this condition because the design of his community inhibits exercising. He might not even have been driving his car if he had been able to bicycle safely to his destination or if public transportation had been available.

This example suggests the myriad ways in which environmental factors are linked with our health on every level. To improve our health we need to adopt a holistic view—one that considers all levels of environmental health at once. We need to address the complexity inherent in environmental health and embrace it. An important way of doing so is to expand the concept of who our environmental health "officials" are. Health promotion and disease prevention are not just the province of physicians, nurses, and public health officials. Urban planners and transportation engineers, landscape architects and developers, local elected officials and mortgage bankers—all of these people, and many others, make decisions that affect the shape of our built environment, the integrity of our natural environment, and the quality of our social environment. Each can contribute in important ways to building a healthier, more sustainable country.

The same principle applies at the level of organizations. Although the Department of Health and Human Services is our main federal health agency, the Environmental Protection Agency, the Department of Transportation, the Department of Energy, the Department of the Interior, and the Department of Housing and Urban Development should also be recognized as health agencies. All of these agencies, and their counterparts at regional, state, and local levels, affect the way we use our natural environment, design our built environment, and behave as individuals and as a community.

WHAT ARE OUR RESEARCH NEEDS?

Traditionally, research in environmental health has focused on toxic exposures. Although research is still needed in this area, we must move beyond environmental toxicology to broader issues of environmental health. We must study the *environmental determinants of behaviors* that affect health. For example, we need to examine our children's exercise patterns and nutritional habits. How much exercise are our children getting, and what impediments to exercising does the urban environment pose? What are our children eating, and what environmental factors contribute to healthy eating patterns and unhealthy ones? Are there environmental determinants of obesity in children? We also need to examine the factors that determine consumer decisions in areas that impact our environment. For example, what determines whether drivers buy large sports utility vehicles or small, ecologically sound, hybrid cars? Probing behavioral issues such as these may yield much information that will potentially change behaviors and benefit environmental health.

We need research on *technical issues*. As in the past, engineering interventions are important potential determinants of public health. A century ago, key environmental determinants of health were sewage treatment and water provision. Today, main determinants include clean energy generation, clean and safe transportation, and the use of industrial ecology to convert the waste stream of one industry into the raw materials of others. Health officials must work with engineers in these areas to develop synergistic research agendas.

We need research on *healthful design and architecture*. Bringing the issue of health into research on architecture, transportation, land use, and resource use will help us make better decisions for creating and restoring our built environment. For example, research is needed on community and regional interventions that will improve opportunities for physical activity, increase the safety of automobile travel, reduce pollution, and develop low-impact environmental materials.

Finally, we need *policy* research. Environmentalists and industrialists have shown great interest in moving beyond the "end-of-pipe" regulatory atmosphere to develop incentives and other innovative methods of changing behavior toward the environment. Research is needed to determine which methods will work, and under what circumstances.

WHERE DO WE GO FROM HERE?

In charting our course as we rebuild the unity between environment and health in the Southeast and throughout the country, we must consider what we need to know, what we need to value, and what we need to do.

What do we need to know? Knowledge is based on research and built through education. We need to educate the public about the multidimensional links between health and environment and how decisions in one area directly affect the other. Educating the public is the key to inspiring environmental health advocacy. The media have an important role in informing the public that, as Paul Rogers has often said, environmental laws are health laws and environmental decisions are health decisions. The more that health issues and environmental issues become fused, the greater will be the progress made in both areas. Several colleges and universities in the Southeast and elsewhere have introduced innovative curricula that emphasize the interdisciplinary nature of environmental education and the importance of environmental literacy for all students. This educational initiative should eventually permeate the primary and secondary educational levels.

What about our values and attitudes? We need to have respect and reverence for the environment, a value that can flourish best when we have respect and reverence for each other. Respect is central to building social capital. We need to value beauty. Appreciation and enjoyment of the beauty of our natural environment should permeate our approach to environment health and form an integral part of all of our policies. We need to value health. The importance of health should be reflected in the behaviors that we adopt and in the policies that we make. Environmentalists should bear in mind that health is a driving force behind the public's involvement in environmental advocacy. We need to value equity and social justice. We cannot advance our environmental agenda, and thus our health agenda, without also embracing equity. Environmental and health practices that protect some communities at the expense of others, or that affect some members of the community disproportionately, must be rectified. Finally, we need to embrace an attitude of collective solutions to problems and collective action to address them.

How about our actions? What do we need to do? We have to join forces across disciplines. Environmental health advocates need to join with architects, planners, engineers, developers, health professionals, epidemiologists, industrialists, members of the clergy, scientists, government officials, activists, policy makers, and the general public. By doing so, we can forge better solutions to environmental health problems; we can also more readily achieve the political majority needed to affect change in environmental and health policies.

We also need to join forces across social boundaries. Environmental and health professionals working in "ivory tower" settings have to reach out to communities and incorporate community decision making into institutional decision making. We as individuals need to reach out to all members of society. As a nation, we must overcome the social desire not to live with different kinds of people, which is reflected in the widespread segregation of housing by class and race. Building communities that attract a mixture of people will move society in this direction.

Several themes have dominated this workshop and will undoubtedly inform the environmental health debate for the foreseeable future. The first is the overarching theme of this conference—that the health of the environment and the health of people are intrinsically intertwined. No longer can we despoil the environment and believe that we, individually or as a community, will not be affected. No longer can we regard human illness as disconnected from the environments in which we live, work, and play. Human health requires that we maintain a healthy environment. In turn, the richness of the environment can enrich our lives in many ways.

The second theme is that traditional methods of ensuring environmental health protection—rules, regulations, and fines—should be supplemented, and in some cases supplanted, by a more enlightened, cooperative approach that brings representatives from many areas together to solve environmental problems as a team. Government regulations should serve mainly as a backstop when other approaches fail. Industry needs to embrace environmental protection, not only because it is the correct thing to do, but also because it will lead to new and more productive ways to do business. Community residents are an immensely powerful force for environmental health advocacy and should not be ignored.

A third theme is the interdisciplinary nature of environmental health efforts. Just as health workers and environmentalists have many common interests, so do housing developers and economists, transportation safety experts and scientists, community residents and auto designers, planners and policy makers. All impact, and are impacted by, the environment, and all belong around the "solution table."

The fourth theme is that technology, although not the whole answer, will certainly be a large part of the solution. Just as several decades ago technological advances enabled unleaded gasoline to be used and greatly reduce environmental lead, future advances will result in other giant steps in environmental protection in this country. We can anticipate some of these changes, because research is already well under way. For example, in 50 to 100 years, energy from the sun, used in high-efficiency photovoltaic cells, may emerge as the predominant power source. Technological advances, coupled with cooperative action in the educational, medical, business, social, and political spheres, offer great hope for protection of environmental health in the Southeast, throughout the United States, and around the world.

1

Perspective on Environmental Health*

Joseph Lowery

This workshop brings together many individuals from diverse fields to consider rebuilding the unity of health and the environment in the southeastern United States. The issues that confront us are far-reaching and complex. The complexity of the interaction of health and the environment is not a new experience for me. I have faced it before in my work in the environmental justice movement.

The term "environmental justice" had its beginning in the early 1980s, when a small community in Warren County, North Carolina, was selected as the site of a massive toxic waste dump. The location of this site in a predominantly black, and overwhelmingly poor, community led to public demonstrations that attracted national media attention and led to hundreds of arrests. As president of the Southern Christian Leadership Conference, I participated in the protest, and I was arrested twice.

During a protest one night, we were marching toward the courthouse, and I looked back to survey the nature of the crowd. Hovering over the crowd was a great cloud of smoke from the many protesters who were smoking. As we marched to protest against the depositing of toxic materials in the ground, the water, and our food sources, we were taking toxins directly into our systems. At that moment, the complexity of the interface between environment and health became startlingly evident.

In that era, 20 or 30 years ago, the nation made substantial progress in addressing issues related to the environment and health, even though scientific knowledge was still rudimentary. Passage of the Safe Drinking Water Act in 1974 and the Clean Air Amendments in 1977 was crucial to the future of environmental health in the United States. Key legislation in 1972 and 1976 required analysis of chemicals to which the public might be exposed through food or other pathways, and the Superfund statute in 1980 addressed hazardous waste disposal.

---

*This chapter is an edited transcript of Dr. Joseph Lowery's remarks at the workshop.

It is a sign of complexity that two or three decades ago, with so little knowledge, so much was done to safeguard health and the environment and that, since that time, with a wealth of knowledge, so little has been done. Twenty years of research has revealed that we, and particularly our children, are vulnerable to environmental injustice. Yet politicians and policy makers have not led the public to see the interrelation between education, environment, and health. They have not led because we have not demanded this leadership from them. A central message of today's meeting is that we must become drivers of those who make public policy and hold them accountable for the condition of the environment.

In a larger sense, we are all responsible for the condition of the earth. The earth is a gift or, more accurately, a lease. A lease implies accountability, and with any lease comes a bill that sets forth the specifics of this accountability. In this view, the earth is leased to us, to be cared for and maintained in good condition for future generations. We face accountability in our stewardship and trusteeship of the earth every day as we use it and take resources from it.

Recently, a radio broadcast reported on the 100 most memorable songs of the twentieth century. One song near the top of the list fits the context of our discussion today: *R-e-s-p-e-c-t,* by Aretha Franklin. As we grapple with the relationship between environment and health, the key is *r-e-s-p-e-c-t*—for the whole creation, for ourselves, and for the environment. In its broadest sense, respect means appreciating and holding in awe the wonders of creation.

The Spanish philosopher Jose Ortega y Gasset expressed beautifully the concept of respect in his words, "I am I plus my surroundings and if I do not preserve the latter, I do not preserve myself" (*Meditations on Quixote*, 1914). These words remind us that the very conditions that we render upon the environment, we render unto ourselves.

In this era, we have become caught up in materialism and greed, and our greed has led us to exploit the poor and deprived. In this country, many are still waiting to gain access to the abundant life they see around them, including health care. The current debate on the patients' rights bill shows some respect for the rights of patients and may lead to legislation, but it does not address the basic issue, which is that 43 million people in this country do not have adequate health care. Patient rights are irrelevant for these 43 million people, because they have no one to sue.

This is not the way to honor the oneness of the human family and to show respect for the creation and the created. Instead, we must move beyond charity to love. Charity gives a hungry man a fish sandwich; love will teach him how to fish—but love will not stop there. Love and respect are interchangeable, and respect means providing training so that a person can get a job, buy his own fish, and buy his fishing equipment. Respect means providing a living wage, health care, and adequate means for retirement. Love and respect for all people, and particularly for the deprived, draw us to strong sense of advocacy.

Our challenge, as we consider the interface between environment and health, is to see our environment and ourselves as one, and to understand that there is no path to fulfillment for any one of us that does not intersect the path to fulfillment for the rest of us. Acknowledging this creative source of interdependence, this brotherhood and sisterhood, compels us to have respect for this creation and for each other. This respect can spur us on to an effective advocacy that moves us from individual concerns to concern for all and that deals not only with effect but with cause.

A story illustrates this point. In a village at the foot of a river, people lived peacefully, in harmony with themselves and the environment. One day a woman saw a baby coming down the river screaming, and she called for help. The men came running, jumped in the water, and saved the baby. They took the child to a warm place and gave him everything he needed.

The next day another baby came down the river. The villagers did the same thing, day after day. Finally, they organized a children's committee; they got the United Way. They did everything they could to take care of the children coming down the river from the mountain day after day. One day, someone said, "I quit. I'm not going to participate in this."

"But you can't! We've still got babies coming!" others protested.

"Yes, I know that. I'm going up the mountain to see who is throwing these babies in the river. I'm going to see if I can't put a stop to it."

That is our advocacy—to fathom the root causes of harm to our environment, and ultimately our health, and to work as one human family to treat those causes and not just their symptoms.

I am pleased that the series of regional workshops on rebuilding the unity of health and the environment has begun here in the Southeast. I believe that the Southeast can lead the nation on this issue. I think that the warmth of our area matches the warmth of our hearts as they flow with respect for the creation.

2

Rebuilding the Unity of Health and the Environment*

Richard J. Jackson

Our discussion of rebuilding the unity of health and the environment in the southeastern United States logically begins with the definition of "health." The World Health Organization defines health as "a state of complete physical, mental and social well-being and not merely the absence of disease or infirmity" (World Health Organization, 1986). Changes in society in the past 100 years have caused us to broaden our definition of health, to expand the role of public health, and to recognize the connection between the environment and health.

Life expectancy in the United States has increased by nearly 30 years in the last 100 years (Centers for Disase Control and Prevention, 1999a). Most of that improvement has come from basic public health measures, such as sanitation, an improved economy, and better housing. Only about seven years are attributed to improved medical care (Bunker et al., 1994). During the same 100 years, the diseases that cause our death have changed dramatically. We seldom die from communicable diseases such as pneumonia, diarrhea, and tuberculosis. We die more often from chronic diseases, such as heart disease, cancer, and lung disease, and from injuries (Centers for Disease Control and Prevention, 1999b). This shift in the major causes of death has presented new challenges for public health agencies. Concepts such as sedentary life-styles, automobile use, diet, and urban sprawl are found increasingly in the vocabulary of public health officials.

During the past 30 years, Americans have benefited greatly from environmental regulations and laws concerning air quality, water quality, waste disposal, and toxic exposure. An example of positive change is the reduced levels of toxic chemicals in the population. The average blood lead level of people in the United States is now 2 µg/dL. Before the passage of the Clean Air Act of the 1970s and the removal of lead from gasoline, the level was 20 µg/dL—enough to reduce IQ scores by 4 to 5 points (Grosse et al., 2002).

Although we welcome these changes, we have in some sense lost touch with

---

*This chapter is an edited transcript of Dr. Richard Jackson's remarks at the workshop.

the vision of what we want our urban environment to be and what quality of life we want to attain. For hundreds of years, people have known how to build urban environments that are dense and also pleasing to human beings—cities in which people feel connected to each other. High urban density is not invariably associated with negative effects on physical or mental health. People enjoy cities with architecturally diverse three- and four-story buildings that encourage and welcome them to walk around—cities such as London and Paris. People enjoy living and working near parkland and cool green spaces and value these natural assets. By contrast, the urban sprawl in the metropolitan areas of our country is characterized by features that detract from the enjoyment of natural and manmade surroundings and reduce the sense of community (Box 2–1).

Box 2–1 What Is Sprawl?

Sprawl is a pattern of urban regional development that features the following:

- Land-extensive, low-density, leapfrog development
- Segregation of land uses
- Extensive road construction
- Architectural homogeneity
- Economic and racial homogeneity
- Shift of development and capital investment from inner cities to the periphery
- Absence of regional planning

What has caused us to diverge so dramatically from the age-old urban design features that were so pleasing in earlier eras? What factors have led to the acceleration of urban sprawl that we are experiencing here in Atlanta and in other U.S. cities? The answers to these questions may help us understand what we can do to modify our design of urban areas, use of natural resources, and lifestyle behaviors to create a healthier and more livable urban environment.

Many forces, including cheap land, technological advances, and social policies, have combined to drive migration from cities to suburbs, creating what some have characterized as a "suburban nation." In Atlanta, the main factor that has influenced urban sprawl is population growth. The population of the Atlanta area has tripled in the past 50 years (Brookings Institution, 2002a), and this rapid growth has placed strains on the natural and human environment. Ominously, the entire U.S. population is expected to more than double in this century, reaching 571 million by the year 2100 (U.S. Census Bureau, 2000c). The environmental issues that may seem remote today will be brought dramatically to the forefront. A burgeoning population is one reason that many cities, including Atlanta, have become very difficult to live in. Commutes have doubled and tripled, and for many people, urban life has become taxing.

> **Box 2-2 How Might Urban Sprawl Affect Health?**
> 1. Increased air pollution
> 2. Increased heat
> 3. Decreased water quality and quantity
> 4. Reduced physical exercise
> 5. More automobile crashes
> 6. More pedestrian injuries
> 7. Mental health consequences
> 8. Decreased social capital

Urban sprawl not only has considerable direct and indirect consequences for the environment, such as loss of forests and depletion of waterways, it also has consequences for human health in at least eight areas (Box 2–2). The first area, air pollution, is a growing health problem in our cities. In Atlanta, high ozone levels are a particular health hazard. Ozone air pollution inflames the airways, affects the immune system, and increases the risk of heart disease and lung disease (Committee of the Environmental and Occupational Health Assembly, 1996). Emergency department admissions nationwide have been shown to increase by 40 percent during ozone alert days (Committee of the Environmental and Occupational Health Assembly, 1996).

Despite the obstacles of rapid population growth and decreasing air quality, the behavioral choices we make can positively affect our environment and our health. For example, to avoid traffic congestion during the Atlanta Summer Olympics in 1996, many people stopped driving and used the city's rapid transit system. The air quality in Atlanta improved by about 30 percent during that time (Friedman et al., 2001). People were in a good mood. Tremendous crowds filled the downtown area. The city was more fun to live in when the air was cleaner, and it was also a healthier city. Children's acute care visits to medical clinics and pediatric emergency departments for asthma decreased, and hospital admissions for respiratory diseases declined throughout the city (Friedman et al., 2001).

Another health hazard posed by urban sprawl is the effect of heat. On warm days, urban areas can be 6 to 8°F warmer than surrounding areas, an effect known as the urban heat island. This effect is caused by two factors. First, dark surfaces, such as roadways and rooftops, efficiently absorb heat from sunlight and reradiate it as thermal infrared radiation; these surfaces can reach temperatures that are 50 to 70°F higher than the surrounding air. Second, urban areas are relatively devoid of vegetation, especially trees, which would provide shade and would cool the air through "evapotranspiration."

In Atlanta, urban sprawl has featured precisely the changes that expand our urban heat island: clearing trees and building large areas of roofs and roadways

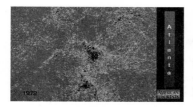 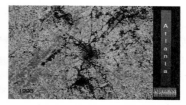

FIGURE 2–1   Atlanta's heat island. Scientists from the National Aeronautics and Space Adiministration have discovered that Atlanta's sprawl development pattern is creating thunderstorms. SOURCE: American Forests, reprinted with permission.

(Figure 2–1). On a warm, 80°F day in Atlanta, the temperature of a concrete airport runway may reach 84°F, an old asphalt road might reach 98°F, and a freshly blacktopped parking lot might reach 102°F (Quattrochi et al., 1998).

The health effects of heat are well known (Nadel and Cullen, 1994). Relatively benign disorders include heat syncope, or fainting; heat edema, or swelling; and heat tetany, a result of heat-induced hyperventilation. Heat cramps are muscle spasms that occur after strenuous exertion in a hot environment, and heat exhaustion is a more severe acute illness.

The most serious condition is heat stroke, which represents a failure of the body to dissipate heat and can be fatal. Heat also has indirect effects on health that are mediated through air pollution. Ozone formation from its precursors, NO_x and hydrocarbons, is enhanced by heat. Also, as heat increases, the demand for energy to power air conditioners rises, requiring power plants to increase their output. The increased demand results in greater production of the pollution that these plants generate, including particulate matter, SO_x, NO_x, and air toxics.

Urban design features can reduce the amount of heat in our cities. For example, a light-colored roof on a home reduces heating and cooling costs by about 15 to 20 percent (U.S. Department of Energy, 2002). Planting vegetation around homes can reduce energy costs by as much as 25 percent (U.S. Department of Energy, 2002). Not only do trees remove CO_2 and produce oxygen, they also cool our environment. Trees and other vegetation slow the runoff of water into streams, allow groundwater recharge, and make the environment more attractive for walking and other physical activities. It is eminently clear that we should safeguard our trees. Yet in the Atlanta region, we have been removing an average of 55 acres of trees every single day, and we have been doing this for 20 years (Quattrochi, 2000).

Removing trees and paving the land have another detrimental effect on health. They diminish water quality. During the first rain after 10 or 12 dry days, the oil, tire rubbings, crushed tire-balancing lead weights, dust, and antifreeze that have accumulated on our asphalt roadways and parking lots are swept into culverts and drains—and ultimately end up in the river. Reducing the amount of runoff in our rivers has genuine health benefits. The Centers for Disease Control and Prevention (CDC) is conducting several studies on the health effects of contaminants in drinking water. Water engineers may tell us that our water meets all requirements for water quality. Yet our water is sampled for less than 10 percent of the carbon materials it contains. The other 90 percent of these materials are unidentified and untested. Our drinking water may meet all current standards, but these standards may not be stringent enough to safeguard our water quality because we tend to evaluate only what we can measure easily.

A recent decision by New York City officials provides an example of an environmentally sound means of improving water treatment. City officials were recently faced with the need to build a $5 billion water treatment plant with an estimated annual operating cost of $200 million for water-purifying chemicals and maintenance. Instead of building the plant, they spent only about $100 million buying buffer land around their main reservoirs in the Catskill region. In so doing, they maintained their water quality for far less cost without requiring much increase in the level of water treatment, and they preserved the watershed lands at the same time.

A further effect of urban sprawl on health is the enormous increase in auto use and the crashes that inevitably follow. In the Atlanta area, we drive 95 million miles each day, enough to drive to the sun and part way back—an average of 36.9 miles for each man, woman, and child in the region, including non-drivers (Lomax et al., 2001). We waste 136 million hours waiting in traffic, or an average of an hour a week for each of us—equivalent to a year's worth of full-time work from 68,000 people (Lomax et al., 2001). We waste 214 million gallons of gasoline, contributing to local and regional air pollution and to global levels of CO_2 (Lomax et al., 2001).

Traffic-related injuries are the leading cause of death among young people in the United States (National Highway Traffic Safety Administration, 1999), and motor vehicle crashes alone account for more than 40,000 deaths a year (CDC, 1999). Automobile fatality rates vary across cities (Table 2–1). In 1998, Atlanta led the list with about 13 deaths per 100,000 people (National Highway Traffic Safety Administration, 1999).

Because driving is so dangerous, we might imagine that people would be safer walking. Statistics prove otherwise. Annual pedestrian fatality rates among major cities in the United States show about 1.9 fatalities per 100,000 people in Philadelphia, 4.6 per 100,000 in San Francisco, and 6.4 per 100,000 in Atlanta (Table 2–1) (National Highway Traffic Safety Administration, 1999). What accounts for the differences? Some cities, such as Atlanta, are tough

---

Our urban environment discourages many forms of beneficial physical activity, as driving replaces walking and bicycling, and as roadways are built without sidewalks, paths, and safe pedestrian crossings.

Richard Jackson

TABLE 2-1 Automobile and Pedestrian Fatality Rates in U.S. Cities

City	Automobile Fatality Rates per year 1998 (deaths per 100,000 people)	Pedestrian Fatality Rates, 1998 (deaths per 100,000 people)
New York	2.51	2.33
San Francisco	3.76	4.55
Philadelphia	5.36	1.88
Portland	6.55	2.58
Houston	9.8	3.41
Phoenix	10.52	4.09
Dallas	11.33	4.28
Atlanta	13.12	6.44

SOURCE: National Highway Traffic Safety Administration, 1999.

towns to walk in. Many areas have no sidewalks and no way to get around on foot. The state of California has addressed this problem by setting a goal that every child ought to be able to walk or bicycle safely to school. Officials hope to achieve this goal by using a set-aside of highway funds to build routes so that children can get to school unharmed.

While increasing our calorie consumption, we have dramatically reduced our physical activity. Being sedentary carries a two- to threefold increase in the risk of early death and a three- to fivefold increase in the risk of dying from heart disease (Wei et al., 1999). The effect of low physical fitness is comparable to that of hypertension, high cholesterol, type II diabetes, and even smoking (Blair et al., 1996; Wei et al., 1999). Conversely, physical activity prolongs life (Lee and Paffenbarger, 2000; Wannamethee et al., 1998), and it also benefits health indirectly, through its effect on body weight.

The United States is currently suffering an epidemic of overweight, which has advanced rapidly in the last two decades. Two-thirds of the U.S. adult population is now considered either overweight or obese (Mokdad et al., 1999). Overweight and obesity are risk factors for a wide range of health problems, including cardiovascular disease and cancer (National Institutes of Health, 1998). Although sprawl does not fully account for our increasingly sedentary lives and our national epidemic of overweight, it is an important contributor to these expanding health problems.

The combination of health hazards in urban areas poses a dilemma for health professionals who counsel urban dwellers on healthy behaviors. For example, how can we promote outdoor physical activities when to do so may place people at risk for exposure to high ozone levels, excessive heat, and pedestrian injuries? Encouraging the pursuit of healthy behaviors in an unhealthy environment sends a mixed message.

Our mental health is another area that is affected by urban sprawl. For example, there is considerable evidence that commuting is linked to back pain, cardiovascular disease, and self-reported stress (Koslowsky et al., 1995). As people spend more time on more crowded roads, an increase in these adverse health outcomes might be expected. One indicator of mental distress related to driving is road rage, defined as "events in which an angry or impatient driver tries to kill or injure another driver after a traffic dispute" (Rathbone and Huckabee, 1999). Road rage appears to be increasing (Mizell, 1997), and the reasons for it are not well understood. Stress at home or work may combine with stress while driving to elicit anger (Harding et al., 1998; Hartley and el Hassani, 1994). Long delays on crowded roads are likely to be a contributing factor. If road rage reflects the stress that accompanies frequent, long, and difficult commutes on crowded roads, it indicates another manner in which sprawl may threaten both mental and physical health.

> A helicopter could drop you at any one of 100,000 intersections, and you would have no idea whether you were in Maine or Virginia, or anywhere else in the United States.
>
> Richard J. Jackson

Exercise preserves both physical health and mental health. Studies show that physical activity has a beneficial effect on symptoms of depression and anxiety and that it improves mood (U.S. Department of Health and Human Services, 1996). Depression is associated with low levels of serotonin. Some studies indicate that higher levels of physical activity can significantly raise serotonin levels and that physical activity is as effective for combating depression as some selective serotonin reuptake inhibitors (SSRIs) that are prescribed for this condition.

Again, our urban environment hinders us from taking part in many forms of beneficial physical activity. For example, the CDC offices in Atlanta are located on a busy highway with no sidewalks. As beneficial to health as bicycling would be, it would be suicidal for a CDC employee to bicycle to work, because doing so requires riding in a gutter next to six lanes of traffic. Not only is this road dangerous, it is ugly. As James Howard Kunstler (1996) writes in his book *Home from Nowhere,* "We drive up and down the gruesome, tragic, suburban boulevards of commerce, and we are overwhelmed by the fantastic, awesome, stupefied ugliness of absolutely everything in sight. It's as if the urban environment has been designed by some diabolical force, bent on making human beings miserable." A helicopter could drop you at any one of 100,000 intersections, and you would have no idea whether you were in Maine or Virginia, or anywhere else in the United States.

We have created a depressing environment that makes us glum about the future of civilization. Although being glum may not strike us as being very serious, we must remind ourselves that the leading chronic disease of American

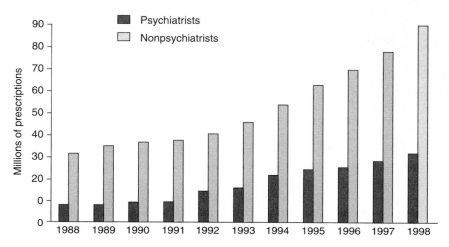

FIGURE 2–2 The use of antidepressant drugs has increased from 1988-1998. SOURCE: Foote and Etheredge (2000). Reprinted with permission from IMS Health, Inc.

adults is depression. Any practicing clinician will tell you that probably half of all medical patients have some concurrent depression or mental health disorder.

The dramatic rise in the consumption of prescription antidepressants over the last ten years suggests that depression, too, has reached epidemic proportions in our society (Figure 2–2). What are the treatments for depression? The obvious answers are taking medication, establishing effective social interactions, and psychotherapy, but pursuing physical activity is also an effective treatment. In cities such as Washington, D.C., and New York, taking a long walk is fairly easy. In Atlanta, we have designed an environment that penalizes people for pursuing physical activity.

A final effect of urban sprawl on health is the diminution of social connectedness, or social capital, which is an accumulation of social networking, civic engagement, and shared trust and reciprocity. Numerous writers have observed a loss of social capital in recent years (Etzioni, 1993; Putnam, 2000), and some authors have attributed this decline, in part, to suburbanization and sprawl (Calthorpe, 1993; Mo and Wilkie, 1997).

How do we approach this set of problems? Those of us in environmental health have spent considerable time over a long period looking at environmental issues in a very narrow way, and at the same time feeling as if the larger environment in which we live is becoming more difficult to control and less connected to human needs. The purpose of this meeting is not so much to focus narrowly on specific issues, such as toxic exposures, but to promote exploration of the larger issues of how the environment influences our total health—physical, mental, and social. Such a holistic effort will enable us to make the connections between health and the environment and to nurture our natural environment, design our built environment, and strengthen our social environment in ways that will promote better health.

3

Environmental Health: A Fifty-Year Perspective*

Jeffrey Koplan

During the past 50 years, substantial progress has been made on environmental health issues. We have progressed from a time when the environment was taken for granted with no thought of its degradation, to a period in which recognition of environmental issues spurred many historic laws, to the present era where we have begun to think about the environment in a much broader interdisciplinary context. This trend might best be illustrated by comparing experiences in two cities, Boston and Atlanta, 50 years ago and today. Boston typifies an older American city—one that experienced most of its suburban growth in the pre-automobile era—and Atlanta exemplifies a modern American city—one that had its greatest suburban expansion in the post-automobile era. In these 50 years, what has changed and what has remained unchanged? What has improved and what has deteriorated?

THE 1950s IN BOSTON

In the 1950s in suburban Boston, clean air was the rule. Air pollution was not a feature of life or even talked about. The water was not fluoridated. Boston was one of the last cities in the country to fluoridate its water, to the detriment of its citizens. In the 1950s, all children in Boston walked to school. The elementary school was about a half a mile away, and the junior high school was a little over a mile away. Many schools had no school buses, and there was no concept of car pooling children to school. There were sidewalks on both sides of every road, and they were considered a necessary part of every community. Tobacco smoke was a feature of life up through the 1970s. Many homes were filled with smoke most of the time. Many parents smoked, and social events were marked by smoke, food, and drink. Smoke was everywhere, in restau-

---

*This chapter is an edited transcript of Dr. Jeffrey Koplan's remarks at the workshop.

rants, in offices, in sports facilities, and even in hospitals. People smoked at the nurses' stations. Following patient rounds, doctors retired to a conference room and everyone pulled out a pipe or a cigar. In that room you could cut the smoke with a knife! When the doctors left the room, a wall of smoke would roll out into the patients' corridor. There was no concept of auto safety. Cars had no seat belts. Cars were for speed, looks, prestige, and transport; safety was rarely considered. These were the features of the environment in the 1950s, 1960s, and even the 1970s, not only in Boston, but in many similar cities throughout the United States.

2001 IN ATLANTA

How have these things changed in a modern city of today—suburban Atlanta? The air quality in Atlanta has deteriorated noticeably from the early 1970s to the present. Haze and pollution are now common features of the city's environment. The water is fluoridated. In fact, cities that do not fluoridate water supplies are now hard to find, and the current threat is the explosive growth of the bottled (unfluoridated) water industry and the concern that children are no longer getting adequate levels of fluoride. Walking is difficult in Atlanta today. Sidewalks are not a feature of most Atlanta communities that were built in the last 30 years. Biking is unsafe in many places because of the explosion of automobile traffic and the absence of bike paths and bike lanes. The lack of physical exercise has resulted in an epidemic of obesity and its sequelae. Most of life is smoke-free, and people are rarely exposed to a smoky environment. Auto safety is now firmly entrenched in the public mind, and lives are being saved as a result. Roads are safer, and cars are safer. Our concept of occupational and environmental safety has matured tremendously. A new aspect of today's life is that workplace stresses follow us around all day. We have difficulty separating work from other parts of our lives because we are constantly accessible by pages, e-mails, faxes, and cell phone calls. This trend is an environmental stressor that is probably not healthy.

If we consider the two locales in the two eras, we see pluses and minuses in each one. For example, although we have become more dependent on cars, we have enhanced auto safety. Although our air has become more polluted, we have reduced our use of tobacco. It is clear that one time period or one city does not represent the ideal. As we improve in some areas of environmental health, we stumble across new hazards that keep us from living healthier lives.

As recently as 25 years ago, issues linking the environment with social injustice, a range of health conditions, or genetic and behavioral factors would not have been discussed. Today we understand that the environment intersects with every element of public health—from combating infectious diseases, to dealing with problems of climate and terrain, to preserving biodiversity. Public health now addresses a wide range of interrelated environmental issues, such as injury control and prevention, toxic exposure, maternal and child health, genetics, physical activity, and obesity. Environment has always been a key factor in public health. It is likely that our appreciation of its importance to public health will continue to grow and that our efforts will improve the environment and health in Atlanta, in Georgia, in the Southeast, and throughout the United States.

4

Human Health and the Natural Environment[*]

The natural environment is the thin layer of life and life supports, called the biosphere, that contains the earth's air, soil, water, and living organisms. The connection between protecting the natural environment and safeguarding human health has been recognized for some time. In recent decades the focus of research and legislation has been identifying and regulating environmental toxics to reduce harmful human exposures. The effect of various environmental exposures, such as toxic chemicals, air pollution, and biological agents on the human body, is commonly perceived as the central problem in environmental health. However, maintaining a healthy environment extends beyond controlling these hazards.

> The effect of various environmental exposures, such as toxic chemicals, air pollution, and biological agents on the human body, is commonly perceived as the central problem in environmental health. However, maintaining a healthy environment extends beyond controlling these hazards.

Preserving the variety of life on earth is also essential to human health. The natural world continually offers compounds that are useful to the pharmacopoeia. Animal and plant products are vital for research and diagnostic tools, and they can be used as indicators of pollution-related disease. Research suggests that biodiversity may hold a key to the prevention and treatment of many diseases (Lovejoy, 2001).

An even more direct connection between the environment and health is the potential enhancement of our physical, mental, and social well-being through our daily exposure to the natural environment. People's nearly universal prefer-

---

[*]This chapter and subsequent chapters were prepared from the transcript of the meeting by Laurie Yelle as the rapporteur.

ence for contact with the natural world—plants, animals, natural landscapes, the sea, and the wilderness—suggests that we as a species may find tranquility in certain natural environments and may derive health benefits from them (Frumkin, 2001). Recent research has confirmed this link. For example, hospitalized postsurgical patients (Ulrich, 1984), employees (Kaplan, 1992), and prisoners (Moore, 1981) have been shown to gain health benefits from exposure to views of nature. Health benefits have also been reported from viewing plants in gardens, interacting with animals (including pets), and participating in wilderness experiences (Frumkin, 2001). This evidence of health benefits from contact with the natural world suggests a broader paradigm of environmental health that includes health-giving environmental exposures (Frumkin, 2001).

A panel of speakers and respondents discussed strategies for ensuring human health through the maintenance of a healthy natural environment. John Sibley, the Georgia Conservancy, noted that in environmental circles the three-legged stool is often used as a metaphor for sustainability. The three "legs" represent the natural world (the environment), the physically built world (the economy), and the social world (equity). Sustainability requires that all three areas be taken into account. Representatives from the three areas must engage in conversation and form partnerships with each other. Sibley noted that the metaphor fails to reflect one essential part of sustainability—the connection between the environment and health. Representatives of the natural environment, the built environment, and the social environment must also work with, and form partnerships with, representatives from the health services community. Sibley invited participants to explore these connections and to consider what new metaphor may be needed to go forward.

VALUING THE NATURAL ENVIRONMENT

Many environmental problems stem from our failure to value the natural environment as we should, according to Eugene Odum, University of Georgia, Institute of Ecology. Current market economics deal largely with human-made goods and services and very little with nature's goods and services (Odum, 1998). The market forces that regulate human-made goods and services in our free market economy are not applied to nature's goods and services because these resources are considered "economic externalities" and are perceived as free. For example, we view clean air and clean waterways as free; even domestic water is so cheap that market forces rarely influence demand. By taking this perspective, however, we fail to appreciate the true costs of these resources. In the past, these "externalities" (for example,

> Only when a natural resource is scarce, as is water in the southwestern United States, is it regarded as having significant value.
>
> *Eugene Odum*

air, water, and the cost of waste treatment) have held little economic concern because the environment seemed large enough to absorb the costs (Odum, 1998). As the human population continues to burgeon and our demands on the environment skyrocket, this assumption will no longer be valid, concluded Odum.

It is important that we understand the actual costs of the goods and services that nature provides. For example, household water bills cover only the expense of pumping, filtering, and delivering water. They do not pay for nature's processing of that water. About a third of the solar energy that reaches our planet is used to conduct the water cycle. The sun evaporates water from the seas, desalinates it in the process, and delivers it via rain clouds to where people need it. If we had to duplicate these services by replacing them with human-made systems, the expense would be extraordinary. Only when a natural resource is scarce, as is water in the southwestern United States, is it regarded as having significant value.

The same analysis extends beyond water and air to resources that grow on the land and lie within the earth. Although we pay for goods that grow (e.g., food and lumber), we do not pay for nature's building up and maintaining the fertility of the soil or the solar energy that makes growth possible. Similarly, we pay for drilling, mining, processing, and transporting the earth's chemical and mineral resources, but not for the effort that nature expended to create them.

As long as natural resources are not regulated by market forces, it is likely that they will not be properly valued. We must find a better way to merge economics and ecology. Is it time to consider the application of market principles as an alternative to environmental regulations? Can we protect the environment in this way? We are used to regulations and have often used them to good effect, but people dislike being regulated, and insufficient attention is paid to 90 percent of existing regulations.

Odum suggested that perhaps market incentives for promoting environmental health and reducing pollution should be considered. Tax relief and other incentives could be used effectively to reward industry for being guardians of the environment. For example, it is expensive for a power company to be a good steward because antipollution equipment is costly to install and operate. If the company passes the cost on to its customers, the price of the power will not be competitive with that offered by the company's less noble competitors. One alternative is to give the company tax relief until the equipment has been paid off. Once all power plants have antipollution equipment, the environment and our health will benefit, and market forces can again take effect.

Extending market forces to environmental resources poses the potential risk of making basic human needs unaffordable for some and thereby increasing social inequity. Although certain changes may raise the price of the basic necessities of life such as water and power, these costs need not be passed on to the poor. The tax system is currently a vehicle for addressing the problems of social inequity, and it could be extended to environmental issues.

The potential benefits of extending market forces to environmental resources are immense. As an example, the state of Georgia in the 1970s assessed the economic value of its coastal marshes at approximately $50,000 an acre, based on the "work" that marshes do to ensure environmental health. As a result, marshes are now considered more valuable left in their natural state than filled in and developed. Odum suggested that a spirited debate about the costs and benefits of extending market principles to environmental health is warranted.

PROTECTING THE NATURAL ENVIRONMENT: LESSONS FROM NATURE

The question of how to eliminate pollution has plagued humans for the last century. Industrial by-products are often difficult to manage in large quantities, and solutions for eliminating waste have often been prohibitively expensive to implement. As a result, the present "solution" is no solution at all: continually dumping waste until there is no place left to put it, except in "someone else's" backyard. In contrast to industrial systems, natural ecosystems are very efficient. Waste is virtually eliminated because it is reused in some productive manner. Source reduction, evident in natural ecosystems, is the ultimate solution to pollution.

Mimicking the workings of natural ecosystems in our industrial complexes would cause raw materials to be used more effectively and waste to be reduced or eliminated. As companies invent ways to mimic nature's efficiency, they benefit from not having to dispose of waste, and they may be able to sell or license the technology for additional profits. When such technology is applied correctly, profits improve, stated Robert Kerr, Georgia Department of Natural Resources.

The current regulatory process generally takes a single-medium view and considers various aspects of pollution and waste control in isolation. Companies may have several environmental permits—an air permit, a wastewater quality discharge permit, and a solid waste permit—but in many cases they have no relationship to each other. Sometimes, for example, companies take the pollutants out of the air and create solid waste, which then must be disposed in a landfill.

A systematic, holistic view is needed to examine the interrelationships in the process of pollution and waste control and to apply them to reduce business and industry's environmental footprint, concluded Kerr. In some cases, several facilities could work together in a cooperative effort. The result would be to transform industrial ecosystems from

> A systematic, holistic view is needed to examine the interrelationships in the process of pollution and waste control and to apply them to reduce business and industry's environmental footprint.
>
> *Robert Kerr*

linear processes that end with waste disposal to a cyclical process more akin to the process that natural ecosystems use to recycle waste. Not only would the impact on the environment be reduced throughout the life cycle of the product in a cost-effective manner, but the environmental ethic would be incorporated into the company's core business philosophy. Such a solution could also potentially transform government regulatory agencies into partners prepared to assist industry in reducing the environmental impact of waste in a cost-effective manner.

This approach has been taken by the Blue Circle Cement Company in Atlanta, which worked with the Pollution Prevention Assistance Division of the Georgia Department of Natural Resources to identify potentially useful waste by-products from other industrial companies in the region. These waste by-products are now used by Blue Circle as raw material or as fuel for making cement. Also, Blue Circle now has the capacity to burn a million used tires as fuel each year, which benefits the environment by reducing air emissions. The company is also looking into using industrial carpet scraps as an additional fuel source—waste that was previously destined for landfills. This effort is only one part of a regional carpet-recycling system being developed by the Department of Natural Resources in concert with Georgia Institute of Technology and the Carpet and Rug Institute. Synergistic methods of waste reduction are also being identified among other industries and organizations.

Working with Georgia Institute of Technology, the Department of Natural Resources has established 18 regional environmental networks throughout the state. The networks hold quarterly meetings in which representatives of various organizations learn from each other and develop relationships so that they can share their waste by-products as raw materials, said Kerr. This effort has extended beyond the manufacturing community to include state prisons, military bases, colleges, and state parks.

Lessons learned from examining the dynamics of natural and industrial ecosystems will better equip environmental agencies to work with industries, businesses, and institutions to reduce their impact on the environment and simultaneously increase profits. The ultimate result will be to minimize public health risks through cost-effective preventive solutions to current waste-generation practices, concluded Kerr.

ENSURING THE HEALTH OF THE NATURAL ENVIRONMENT: POTENTIAL STRATEGIES

A prevailing theme among conservationists has been that preserving nature and protecting natural areas require keeping them pristine and completely free of the imprints of humans and human systems. This view is in many ways no longer practical because most ecosystems today are impacted in some way by human behavior, stated Matthew Kales, Upper Chattahoochee Riverkeeper. Virtually every stream in the world is affected by atmospheric deposition. The air

> **Box 4–1 Ways to Protect the Natural Environment**
>
> 1. Monitor the health of our local environment actively and continuously
> 2. Create outreach programs for educating individuals about environmental health issues
> 3. Continue to address issues related to pollution
> 4. Base policy about the environment and health on sound science
> 5. Strategies of improving environmental health need to include the particular circumstances of each locality

quality in some of our national parks has been found to be no better than in some of our cities. Essentially, no place exists where we cannot feel, in some measurable way, the footprint of humans. All solutions to environmental health problems must be grounded in this reality. Any proposed solution to problems in the natural environment that discounts the impact of the social and the built environments will be inadequate. To protect the natural environment, solutions are needed that consider the entire environment in a holistic way (Box 4–1).

A first step is to monitor the health of our local environment actively and continuously, said many participants. A set of indices for the health of the environment (e.g., rate of biomass production and respiration, microorganism activity, rate of erosion, levels of toxins) would create a profile of a healthy environment and serve as important benchmarks against which to compare future changes in the environment, noted Odum.

A second step is to create outreach programs for educating individuals about environmental health issues such as water quality. An example of such a program is the bacteria alert network for the Chattahoochee River conducted by Upper Chattahoochee Riverkeeper in concert with the Georgia Conservancy and several federal and state agencies, reported Kales. The Chattahoochee River is a prime resource for the area, supporting navigation and hydropower, providing drinking water, assimilating wastewater, and providing a rich environment for many recreational activities—fishing, boating, swimming, paddling, and walking. Readings of *Escherichia coli* and other bacteria harmful to human health have recently been found to be extremely high. Representatives from Riverkeeper and the National Park Service are taking water samples and publicizing the condition of the river to let the public know whether the area is safe for recreation. A related program is one that offers outreach to "subsistence anglers," people who fish for food, to inform them when bacterial counts indicate that the fish are not safe to eat. In this instance, merely publishing passing guidelines is inadequate. Materials must be available in forms that will reach all affected individuals, perhaps in pictorial form or in languages other than English.

A third step is to continue to address issues related to pollution. Extensive networks and partnerships among industries and between government and industry must be created to reduce waste by-products and minimize the health effects of pollution. Fourth, our decisions about the environment need to be based on sound science, stated many participants. Fifth, approaches to environmental health, including generating environmental indices, have to take into account the particular circumstances of each locality, suggested Samuel Wilson, National Institute of Environmental Health Sciences, National Institutes of Health. Strategies that are the most effective may be different in the Southeast than in other regions, said Wilson. Many participants agreed that the local community must work as a unit to define local environmental problems, to generate creative solutions, and to advocate the adoption of those solutions.

5

Human Health and the Built Environment

In the United States, the "built" environment—the environment designed and constructed by humans—has been greatly influenced by the rapid population growth in this country in the last 50 years. During that time, the U.S. population has increased by 83 percent, and a result is an enormous growth in the number of large metropolitan areas throughout the country (U.S. Census Bureau, 2000b). Land area has expanded much faster than population in many metropolitan areas, demonstrating the land-intensive nature of this pattern of growth. Similarly, the population outside the central cities has grown faster than the population in the central cities, demonstrating a shift of population to suburban areas.

From 1990 to 1998, the metropolitan population in the South grew by 5.3 percent inside the central cities and 18.4 percent outside the central cities (U.S. Census Bureau, 2000b). Metropolitan Atlanta's population grew rapidly from 1970 to 1999, increasing by 114 percent throughout the entire region, but decreasing by 14 percent within the city's borders. The area of metropolitan Atlanta has also grown, from "only" 65 miles from north to south in 1990 to its current size of 110 miles. This pattern of sprawl, typical of the Sunbelt cities, results in low population density across metropolitan areas. Older cities such as Washington, D.C., and Boston tend to have higher density (3,465 and 2,610 people per square mile, respectively), whereas the density of Atlanta and Dallas is only 1,400 people per square mile (Brookings Institute, 2002a).

Rapid population growth is a worldwide phenomenon. In another 50 years, by 2050, about 10 billion people will share this planet, compared with about 6 billion today. About 7.5 billion people will be living in urban areas. Many will live in megacities with populations of more than 10 million people. Many participants agreed that this burgeoning of the population worldwide is a central driving force in environmental problems and has serious ramifications for health that will worsen in the next few decades. Some anticipated consequences are water shortages, increased air and water pollution, overcrowding, increased sprawl and traffic congestion, deforestation and soil erosion, vanishing open spaces, and destruction of wildlife habitats.

A panel of speakers and respondents discussed approaches to making the built environment safer, healthier, and more environmentally friendly. Many participants acknowledged that, particularly in light of expected population growth, the decisions made today, and the solutions devised and implemented in the coming decade, will have profound effects on our environment and the health of the human family generation after generation.

TRANSPORTATION AND HEALTHY ENVIRONMENTS

A close relationship exists between transportation, the design of the built environment, and health, according to Ricardo Martinez, Safety Intelligence Systems.* Modern transportation, particularly the automobile, has become a vital resource for our society and our economy and has given us tremendous mobility. At the same time, it has changed every aspect of our lives—how we behave, how we build houses, how we shape our communities, how we work, and how we play—everything. In the United States, the automobile is among our greatest health hazards, particularly for young people. Automobiles now claim more than 40,000 lives each year in the United States (National Highway Traffic Safety Administration, 1999) and are the leading cause of death among persons 1 to 24 years of age (CDC, 1999b). These fatalities affect African Americans disproportionately. Among men, national rates are highest for African Americans (32.5 deaths per 100,000 people per year), next highest for whites (19.5), and lowest for Hispanics (10.2). Among women, the rates are highest for African Americans (11.6) and roughly equivalent for whites and Hispanics (8.5 and 9.1 per 100,000, respectively) (Cubbin et al., 2000).

> Yet too often we consider transportation only in terms of the freedom it provides and fail to see how closely it is linked with the environment, public health, and safety.
>
> *Ricardo Martinez*

Automobile crashes also account for 3.4 million nonfatal injuries each year in this country at an estimated cost of $200 billion (CDC, 1999b). The rates of automobile fatalities and injuries per driver, and per miles driven, have decreased substantially in recent decades as a result of safer cars and roads, laws that discourage drunk driving, and other measures; yet because of the increased number of cars and miles driven, the absolute toll from automobile crashes remains high.

The automobile also poses a serious health risk for pedestrians in our soci-

---

*Former National Highway Traffic Safety Administration administrator.

FIGURE 5-1   Pedestrians are at risk of injury and fatality from automobiles. SOURCE: Richard J. Jackson. Reprinted with permission.

ety (Figure 5-1). Each year, automobiles cause about 6,000 fatalities and about 110,000 injuries among pedestrians nationwide (Cohen et al., 1997; McCann and DeLille, 2000). These fatalities disproportionately affect members of minority groups. In Atlanta, for instance, annual pedestrian fatality rates from 1994 to 1998 were 9.74 per 100,000 for Hispanics, 3.85 per 100,000 for African Americans, and 1.64 per 100,000 for whites (Hanzlick et al., 1999). Similar patterns are seen across the country (Marosi, 1999; Morano and Sipress, 1999). The reasons for this disproportionate impact are complex and may involve the probability of being a pedestrian (perhaps related to low access to automobiles and public transportation), road construction in areas where members of minority groups walk, and behavioral and cultural factors such as being unaccustomed to high-speed traffic.

An automobile injury or an auto-pedestrian accident can be considered a disease process, suggested Martinez. Like any disease process, it can be separated into three parts: host, agent, and environment. The host is the human, the agent is the energy transmitted to the body (beyond what the tissues can tolerate), and the environment is what brings the two together in the crash. The design of the automobile environment can help keep host and agent—human and energy—apart and greatly increase safety.

Environmental design is an integral part of automobile injury prevention. For example, controlled-access freeways and turnpikes have the lowest fatality rates of our roadways, even though cars travel at high speeds, because deadly head-on and side-impact crashes are avoided (Martinez, 1990).

Interstate highways are designed to provide a relatively safe environment for the automobile. Traffic proceeds in one direction, and there is no cross-traffic. These roadways have guard rails, breakaway poles, wide shoulders, and few distractions. In contrast, the typical state highway has oncoming and cross-traffic, narrow shoulders, many distractions—and higher fatality rates. Incorporating safety features, such as broad median strips and wide shoulders, into the design of state highways can potentially lessen injury rates. "Traffic-calming" design features, such as narrowing roadways to reduce automobiles speeds, may also decrease injuries.

Martinez suggested that proper environmental design can also protect pedestrians. Because a pedestrian always loses in a contest with an automobile, a primary rule is to keep the two separated with sidewalks, overpasses, fences, and pedestrian-friendly intersections. Pedestrian routes that need special attention are those within a community, particularly the routes that children take to reach each other's homes, their school, or a nearby park. Safe pedestrian routes that connect communities with one another are also needed, as are routes leading to and from public transportation and the buildings in which we work and shop.

A serious impediment to pedestrian safety is the absence of sidewalks in many developments in suburban America. Requiring sidewalks in new developments and retrofitting sidewalks in existing communities are possible solutions. Providing incentives for sidewalk construction may be an effective way to bring about community cooperation, suggested Lawrence Frank, Georgia Institute of Technology.

Safety education and traffic law enforcement are important ways of changing driving and pedestrian behavior and promoting safety. Yet the design of the transportation environment can have an even greater effect on safety because it is a preventive measure that does not rely on the public's compliance with safety measures, stated Martinez. The federal government can assist in the design of a safer transportation environment and in funding construction costs. Regional and state approaches to transportation design may provide a unifying perspective. However, much of the work to be done is dictated by the particular circumstances of the locality and often must be addressed at the local level.

THE BUILT ENVIRONMENT AND HEALTH PROBLEMS

Reliance on the automobile, and the environmental and behavioral changes that accompany automobile use, have contributed to many serious public health problems in this country, according to Wayne Alexander, Emory University School of Medicine. The growth of our urban areas has led to decreased physical activity as driving has replaced walking and bicycling. A sedentary life-style, in turn, is a well-established risk factor for cardiovascular disease, stroke, and all-cause mortality (NIH Consensus Development Panel on Physical Activity and Cardiovascular Health, 1996; Pate et al., 1995; U.S. Department of Health and Human Services, 1996; Wannamethee et al., 1998, 1999). Men in the lowest quintile of physical fitness have a two- to threefold increased risk of dying overall, and a three- to fivefold increased risk of dying of cardiovascular disease, compared with men who are more fit (Wei et al., 1999).

Physical activity prolongs life (Lee and Paffenbarger, 2000; Wannamethee et al., 1998). Among women, walking 10 blocks per day or more is associated with a 33 percent decrease in the risk of cardiovascular disease (Sesso et al., 1999). Many chronic diseases involve the cardiovascular system, and it is now understood at the cellular and molecular level why stimulating blood flow in vessels by walking contributes to the health of the cardiovascular system. From the perspective of cardiovascular health alone, an environment that hinders walking and other forms of exercise is an unhealthy one, concluded Alexander.

Obesity resulting from overeating and underexercising is an epidemic that is growing at an alarming rate in this country. Jackson stated that the increase in obesity in this country during the past decade is dramatic (Figure 5–2). The average 11-year-old boy in the United States today is 11 pounds heavier than a boy of the same age in 1973 (CDC, 1999a). Being overweight increases the overall risk of death by 250 percent; it carries a fourfold increase in the risk of heart disease and death and a fivefold increase in the risk of type II diabetes (Willet et al., 1999). Obesity also increases the risk of high blood pressure, gallbladder disease, and some cancers (NIH, 1998).

According to Alexander, air pollution is another serious health consequence of our dependence on the automobile. Air pollution is an umbrella term for a series of distinct contaminants that may be found in air. In general, there are nine kinds of air pollutants: ozone, NO_x, CO, particulates, hydrocarbons, lead, SO_x, air toxics, and allergens. The first five of them are produced by so-called mobile sources—cars and trucks—noted Jackson. In Atlanta, ozone is the leading air pollutant of concern (Table 5–1). Because ozone is a secondary product of hydrocarbons and NO_x, it demonstrates a characteristic daily pattern. On a typical August day in Atlanta, the day starts with very low ozone levels and with a busy

TABLE 5–1 Atlanta's Leading Pollutant Concern: Ozone

Respiratory Effects (ozone > PM)	Cardiovascular Effects (PM > ozone)	Immune Effects
Airway inflammation	Increased mortality	Increased susceptibility to infection
Decreased air flow		
Increased symptoms, emergency department visits, medication use, hospitalizations		

NOTE: PM = particulate matter.

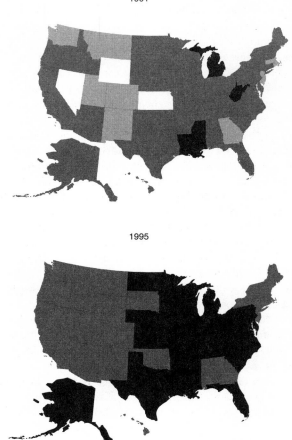

FIGURE 5-2 Prevalence of obesity among U.S. adults: 1991, 1993, 1995, and 1998. Obesity defined as >30 kg/m² body mass index. Data from the CDC's Behavioral Risk Factor Surveillance System. SOURCE: Mokdad et al., 1999. Reprinted with permission.

rush hour. The second rush hour occurs in late afternoon, fueling the process, so that by the time school athletes are outside practicing, bicycle commuters are peddling home, and afternoon joggers are getting their exercise, the air is dangerous to breathe.

Air pollution, like so many health hazards, does not affect everyone in society equally. Poor people and people of color are disproportionately affected for two reasons: (1) they are disproportionately exposed, and (2) they are more

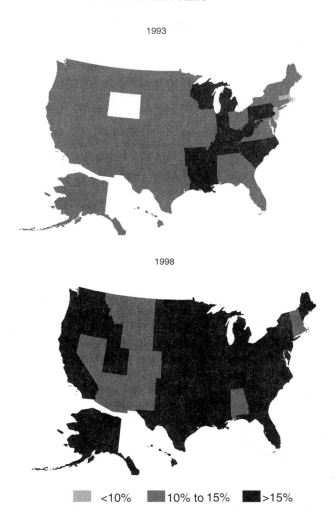

likely to have underlying diseases that increase their susceptibility (Wernett and Nieves, 1992).

There is compelling evidence that air pollution contributes to chronic lung disease and asthma, said Alexander. The epidemic of asthma is growing in our country and is of particular concern in children. As asthma continues to increase, its prevalence and mortality remain higher in minority group members than in whites (Litonjua et al., 1999; Metzger et al., 1995; National Heart, Lung, and Blood Institute Working Group, 1995; Persky et al., 1998).

An important function of health professionals and environmentalists is to educate the public about the strong links between unhealthy life-styles and the environments that promote them. Kerr noted that although protecting the environment against unhealthy influences is difficult and expensive, it may be *more* expensive to deal with the consequences of *not* protecting the environment, particularly increased medical costs. Odum further reflected that an unhealthy environment not only is responsible for disease, but also affects the social health and well-being of the entire community—what the Constitution calls "domestic tranquillity."

ENVIRONMENTALLY FRIENDLY BUILDINGS

The "sick building syndrome" is familiar to many and represents inadequate building design and construction practices. However, building design and construction need not adversely affect the environment or our health. The Southface Energy Institute of Georgia is a research and educational organization that, among other things, promotes environmentally friendly buildings that bring health benefits to their occupants. The Southface building, which serves as a model to reflect the organization's principles, was designed with four goals in mind, according to Dennis Creech, Southface Energy Institute: to promote health, to conserve energy, to improve water efficiency, and to demonstrate the use of low-impact environmental materials (Figure 5–3).

In the Southface building, moisture is controlled to avoid mold and dust mites that can cause health problems such as asthma and allergies. Materials are chosen wisely to avoid substances that can affect health adversely; for example, all paints used in the building contain no volatile organic compounds.

The Southface building reduces the impact of one of our most environmentally damaging activities—energy production and use—by using about 60 percent less energy than the typical building in Atlanta. This energy efficiency is achieved through sound building and insulating techniques and the use of solar electric roof shingles.

Water efficiency is achieved by using less water and preserving water quality. For example, the driveways adjacent to the Southface building are constructed from porous concrete, which allows the rainwater to percolate into the soil to help protect the watershed. Throughout three years of drought, a butterfly garden next to the building has been irrigated with gray water from the laundry.

Low-impact environmental materials are those that require minimal natural resources. The entire shell of the Southface building—the roof and outer walls—is made with insulating panels that use 25 percent less wood than standard construction methods. These materials not only reduce the burden on our forests, but also provide better insulation.

FIGURE 5-3 The Southface building was built to promote health, conserve energy, improve water efficiency and demonstrate the use of low-impact environmental materials. SOURCE: Neal Dent. Reprinted with permission from Southface Energy Institute.

The lessons learned from the Southface building, and others like it, can be applied directly to residential housing. Homes built according to these principles not only save their owners money in energy costs and maintenance—typically a 30 percent saving over a traditional home—but also cause fewer allergy problems, stated Creech.

Many traditional homes are energy inefficient, as can be determined by examining heat loss (Figure 5-4). Windows are a major source of energy loss, but loss from other areas may also be substantial and can be more easily and inexpensively corrected. Caulking wall-to-floor joints and around window and door frames is inexpensive and can make a large difference in stemming energy losses (Georgia Environmental Facilities Authority, 1999). Another area of substantial heat loss is heating ducts. In a typical home, 10 to 30 percent of the

FIGURE 5-4 Infrared image of heat loss. SOURCE: Dennis Creech. Reprinted with permission from Southface Energy Institute.

heating bill is attributable to leaks in the ducts (Georgia Environmental Facilities Authority, 1999).

Creech noted that energy losses not only inflate our heating bills but also cause health problems. When energy escapes through a leak, that area of the home cools down, creating condensation. The moisture encourages the growth of toxic mold and dust mites, which cause allergies in many people. Leaky heat ducts create pressure imbalances in the home that cause cold air to be drawn in from the crawl space underneath the home—a space that is ripe with mold and other biological contaminants. Fixing leaks in heating ducts not only conserves resources but also enhances health.

A tightly sealed house requires that homeowners take care with combustion appliances and cars, both of which can produce lethal levels of carbon monoxide. Unvented combustion appliances have no place in the home. The biggest health risk from combustion in the home is not the water heater, the furnace, or the fireplace—it is the automobile, suggested Creech. In many homes, the leakiest wall is between the home and the attached garage. For that reason alone, the car's engine should never be run inside the garage. Any wind that blows into the open garage when the car's engine is running carries carbon monoxide into the house through the smallest crack. Installing carbon monoxide detectors in all homes that have attached garages would minimize the risk of carbon monoxide poisoning.

PARTNERSHIPS WITH ACADEMIA

Fifty years ago, a research scientist's work typically involved investigating a specific topic in isolation in a university laboratory. The work was generally conducted within a single discipline and was unconnected to other disciplines and to the needs of society. Today's research is often conducted along interdisciplinary lines, and the university now functions as an integral part of the community, making connections with other research organizations and with society as a whole, according to Charles Liotta, Georgia Institute of Technology.

Scientists, engineers, and architects must become health and environment leaders and decision makers rather than just technical problem solvers. Liotta noted that achieving this goal requires new kinds of partnerships. Universities must adopt a more comprehensive view of how students and faculty learn and how they conduct research, and they must promote the changes necessary for scientists, engineers, and architects to be able to create more sustainable technologies.

> Research universities must become a major intellectual force in rebuilding the unity of health and the environment through a holistic approach, rather than through fragmented initiatives.
>
> *Charles Liotta*

One way of promoting sustainable technologies is through partnerships—among departments within universities, between universities and other academic research institutions, and between these institutions and industry, federal laboratories, and other government organizations, suggested Liotta. An example of a useful collaborative technology that is currently being developed is a new type of photovoltaic cell that will allow relatively high efficiency conversion of sunlight into electricity at little cost.

Georgia Institute of Technology has fostered multidisciplinary research by locating centers representing separate but related disciplines in close proximity to each other. For example, the Georgia Center for Advanced Telecommunications Technologies and the Micro-electronics Research Center collocate science, engineering, and social science disciplines so that collaborative efforts can be undertaken to solve real-world problems. Such partnerships create a holistic approach to teaching and learning by giving students and faculty exposure to many different disciplines while furthering in-depth study in their primary discipline.

PARTNERSHIPS WITH INDUSTRY: CREATING TRUST

The existing model of government-generated environmental regulation is "command and control." The goal is to control "end-of-pipe problems" by commanding and enforcing regulations, according to Linda DiSantis, United Parcel Service. Until recently, much of the regulatory work on environmental impacts has been necessary, and the regulatory and compliance process has generally worked well for most industries. Today, however, new kinds of challenges are arising from different kinds of industries and different kinds of impacts, and the traditional model of environmental regulation no longer always works well for the service sector. An answer is to seek creative solutions, such as incentives, for reducing unhealthful impacts on the environment. All major players in the process—government regulators, business and industry, environmental groups, and the research community—appear to favor a holistic approach to improving environmental quality, but a key impediment is the lack of trust among these groups. Although creative solutions are often sought in negotiations between representatives from government and from business and industry, government regulators often revert to the old approach—rule making—primarily because they do not trust business and industry to do the right thing unless penalties are threatened, according to DiSantis. For the holistic approach to work, participating groups must develop mutual trust, and old patterns of behavior must be put aside. Regulators need to reserve rule making for instances when other avenues do not work. Environmental groups play an important part in raising awareness of environmental problems and holding business and government accountable for the role each is supposed to play in solving them.

> Lack of trust amongst major players creates obstacles to achieving a consensus on needed solutions.
>
> Linda DiSantis

Some industries often find themselves dealing with two juxtaposed concerns: (1) economic realities and the ability of companies to deliver a fair return on investment, on one hand, and (2) environmental protection and minimization of health risks, on the other, according to Charles Goodman, of Southern Company. Such is the case with implementing the Clean Air Act, which was a milestone piece of environmental legislation that affected virtually every business—from large energy companies to the local dry cleaners. Most large utilities are significantly affected by clean air regulations, and they recognize the need to control emissions and waste. These companies must deal with the challenge of complying with regulations of all kinds, meeting customer needs, and at the same time satisfying their investors. Most electricity sales are still subject to price regulation, and this means that utilities must deal with some regulators who determine what they can charge for their product and other regulators who try to limit the effects of energy production on the environment. Electricity that is sold on the open market, like other products not subject to price controls, must still face competitive pressure that can limit recovery of the full cost of additional environmental requirements.

Goodman noted that industries may have to incur significant expenses and modify their operations to meet federal standards for air quality, but from an economic vantage point, industries may question what represents good enough. Achieving the last few percentage points of improvement in air quality costs disproportionately more than obtaining the first percentage point. While acknowledging that there are no easy answers to the question, "How clean is clean enough?" some participants suggested that having an increased dialogue would be helpful.

BUILDING HEALTHIER CITIES

The "American dream" has always been to have a good home, a good job, good education, and a safe neighborhood. Achieving this dream in today's society typically means living in the suburbs and relying on the private automobile to drive long distances to work, shop, and play. What are the costs of the pursuit of this goal for individuals, society, and environmental quality? They are well-known—urban sprawl, traffic congestion, long commute times, lack of open space, stress, lack of a sense of community, and poor air quality, according to Michael Kilgallon, the Pacific Group. Atlanta is one of the least densely developed (i.e., fastest-sprawling) cities in the United States. Atlanta's residential lots are, on average, three times the size of lots in any other large city in the South. As the Atlanta area grows by an anticipated 600,000 people in the next decade, continuing to achieve the American dream will mean greatly increasing sprawl and its ill effects. Allowing our society to grow in this manner places us in danger of creating an environment that we will want to "get through" as opposed to an environment that we will enjoy, said Lawrence Frank, Georgia Institute of Technology.

What new direction can we take to accommodate this growth? Across the country, many cities and counties are exploring new policies that would reconfigure urban growth by aiming for higher-density, mixed use development in some areas and preservation of green space in others. Many in the housing industry, in

> **Box 5–1 Regional Development and Smart Growth Features**
>
> A desirable pattern of regional development:
> - Protects and improves the quality of life for all citizens,
> - Permits and promotes healthy behaviors,
> - Minimizes or eliminates hazards to people, and
> - Protects, preserves, and restores the natural environment.
>
> Smart growth is likely to feature the following:
> - Higher-density, more contiguous development
> - Preserved green spaces
> - Mixed land uses with walkable neighborhoods
> - Limited road construction balanced by transportation alternatives
> - Architectural heterogeneity
> - Economic and racial heterogeneity
> - Development and capital investment balanced between central city and periphery
> - Effective, coordinated regional planning

government, and in environmental groups are promoting the concept of "smart growth" as a better way for society to grow (Box 5–1). From a health standpoint, the features of smart growth are highly attractive. They offer the potential of more walking, less driving, fewer car crashes and pedestrian fatalities, and more livable regions with enhanced social capital—all important health benefits.

To improve the built environment, we need to understand the process that shapes it, said Elliott Sclar, Columbia University. Urban planners, engineers, and developers face community barriers, political barriers, and regulatory barriers, stated Kilgallon. Because political and regulatory barriers flow from the community, a useful course of action is to focus on community barriers. Sclar noted that too often, community policy reflects the short-term needs of the few, not the long-term needs of the many. However, as much as most people oppose the effects of sprawl, they also tend to oppose changes that increase density, stated Frank. As is often said, people hate density, and they hate sprawl. The probable solution is to find a middle ground.

A means of combating sprawl that is advocated by many people, although they tend not to specify how to achieve it, is to slow growth or halt it altogether. Kilgallon noted that the only way to slow or halt growth voluntarily in an area is to let the quality of life deteriorate to the point that people no longer want to live there. An alternative, proposed by some participants, is to convince local politicians to stop the growth of existing communities through regulation. However, Kilgallon asserted that halting growth in some communities may merely push the growth out further.

How do we facilitate the adoption of middle-ground concepts such as smart growth in our communities? We must educate each other and the general public about their benefits to health and well-being and convince people of their worth. Many groups and partnerships—among industry, environmental groups, regional governments, and state government—are working toward this goal. Yet because the involvement of local government is missing in most of them, progress is slow. The goal of building healthier cities must be achieved through consensus because voters will not allow regulations to be forced upon them. Thus, the challenge is to show the public the deep connection between smart-growth concepts and a healthier, more productive, and more enjoyable way of living.

6

Human Health and the Social Environment

The social environment can be defined broadly as the social conditions that influence our lives and the life of our communities. An important part of the social environment is our social connectedness, which ranges from our individual interactions with one another to interactions in groups and organizations. The concept that social connectedness is essential for ensuring human health and well-being has been the subject of much research and is now well established (see Putnam, 2000, for review).

A panel of speakers and respondents explored the concept of social capital, delineated the positive effects of social capital on health, and discussed means of restoring social capital in our communities. They also described the tenets of the environmental justice movement, discussed areas in which environmental justice is still needed, and emphasized the importance of bringing environmental justice to bear in every part of society.

SOCIAL CAPITAL

Economists often define accumulated items of worth as capital, a limited resource that enhances productivity. The term "capital" can refer to physical objects (tools, materials, structures), human properties (individual intellect, education, training), or social connections within a community (social networks, civic engagement). Whereas human capital refers only to the contribution of the individual, social capital can be defined as the accumulated connections, or social networks, among individuals and the norms of shared trustworthiness and reciprocity that arise from them (Putnam, 2000). The term "*social* capital" calls attention to the ways in which our lives are made more productive by social ties (Putnam, 2000).

Social connections, or networks, are formed informally among family and friends and in neighborhoods, and more formally in the workplace, in places of worship, and in a variety of other organizations. Social capital implies a continuity of interaction over time and a level of trust that allows people to feel comfortable working together to accomplish goals for the common good. Social capital includes specific reciprocity between individuals and a norm of generalized reciprocity, where something is done with no expectation of immediate return but with the expectation that the return will come later from elsewhere (Putnam, 2000). In short, social capital is the "glue" that holds society together.

Creating and maintaining social capital are ways of ensuring that people stay healthy and that communities are healthy environments, stated Winsome Hawkins, the Community Foundation for Greater Atlanta. The positive effects of social capital on various areas of individual and community life have been well documented (Knack and Keefer, 1997; La Porta et al., 2000). Economic studies have shown that increased social capital makes workers more productive, businesses more effective, and nations more prosperous (Putnam, 2000). Studies indicate that high levels of social capital make individuals less prone to depression and more inclined to help each other and that social capital decreases the rate of suicide and the incidence of colds, heart disease, strokes, and cancer (Putnam, 2000). Research even suggests that joining just one group boosts life expectancy as much as if a smoker stops smoking. Sociological studies reveal that crime, juvenile delinquency, teen pregnancy, and child abuse decrease in areas of increased social capital. Political studies show that increased social capital makes government agencies more responsive, efficient, and innovative. On an individual level, social capital ensures that if we are sick or unemployed, we will receive help from others and that we will have companionship as we enjoy the benefits of our community (Putnam, 2000).

> Creating and maintaining social capital are ways of ensuring that people stay healthy and that communities are healthy environments.
>
> *Winsome Hawkins*

Throughout America's history, levels of civic engagement have risen and fallen (Putnam, 2000). In the last 25 years, levels of social capital appear to have declined in this country (Etzioni, 1993; Putnam, 2000). A survey was recently undertaken by 40 community foundations across the nation to establish benchmark levels of social capital within their respective geographic areas (Saguaro Seminar, 2002). The study measured levels of trust, diversity and friendship, political participation, informal socializing, civic and associational involvement, giving and volunteering, and faith-based engagement. On the index of giving and volunteering, respondents in metropolitan Atlanta scored higher than the national level (39 percent versus 35 percent, respectively) (Horne, 2001), but they scored lower than the national level on social trust (24 percent versus 33 percent, respectively). Only 19 percent of metropolitan Atlanta respondents reported a high level of participation in civic activity compared with the national level of 24 percent. Hawkin noted that a comparison of social capital and indicators of child well-being by area of the country showed that social capital was low in areas with few indicators of child well-being.

These findings suggest that levels of social capital need to be restored in the Southeast and throughout society. Where levels of trust are low, we must find ways to encourage trust and to promote the perception that individual efforts contribute to the good of the whole.

Community building and social capital formation are needed to move our society forward, said Elliot Sclar. A master's program in community building has recently been established at Columbia University, he noted. The program aims to teach students methods for strengthening the social capital of communities. These methods involve commanding resources and engaging in the political process of policy change in areas such as land use and transportation.

> Community building and social capital formation are needed to move our society forward.
>
> Elliot Sclar

Educators need to ensure that information on community building is understandable so that all community members are inspired to engage in restoring social capital, stated Hawkins. A great deal of social capital already exists in faith communities throughout the Southeast, she noted. Building unity among these smaller community groups by bringing them together around broader issues may be an effective way to increase social capital on a larger scale.

ENVIRONMENTAL JUSTICE

The environmental justice movement is a multiethnic, multicultural movement that was founded in the South in the early 1980s as a response to the placement of a toxic waste site in a predominantly poor, minority community (Bullard, 2000). The movement embraces the principle that all communities have a right to equal protection under the laws and regulations governing the environment, housing, transportation, and civil rights. Environmental justice ensures that no population, because of policy or economic disempowerment, is forced to bear a disproportionate burden of the negative human health or environmental impacts of pollution or other environmental consequences (Bullard, 1996).

The impetus for the environmental justice movement has come from grassroots community groups that have forced the issue to the forefront of the nation's attention, stated Robert Bullard, Clark Atlanta University. Through coalitions, alliances, and partnerships, these organizations have battled to keep the issues of disparity in health, environment, transportation, and housing visible to govern-

> The first and foremost principle, is that people must speak for themselves and that their opinion must be respected; agreement is not necessary, but respect must maintained.
>
> Robert Bullard

ment officials and to the public. Seventeen principles of environmental justice were developed at a summit in Washington, D.C., in 1991 (Lee, 1992). This principle safeguards the right of all community members, especially those who are unequally affected by environmental hazards, to participate in the dialogue that advocates change.

In Atlanta, many issues involving environmental justice still need to be addressed. Large disparities among racial and ethnic groups continue to exist in land use patterns, housing, transportation, air quality, and toxic exposures. Nearly 83 percent of Atlanta's African-American population compared to 60 percent of whites lives in zip codes that have an uncontrolled hazardous waste site. While African Americans and other minorities constitute 29.8 percent of the population in the five most populous counties that are contiguous to Atlanta (Fulton, Cobb, DeKalb, Gwinnett, and Clayton counties), they represent the majority of residents in five of the ten "dirtiest" zip codes in these large counties (Bullard et al., 2000). Nationally, 57 percent of whites, 65 percent of African Americans, and 80 percent of Hispanics live in areas with substandard air quality (Wernett and Nieves, 1992). The poor air quality has a disproportionate impact on the health of poor children, poor adults, and people of color. Transportation-related air pollution has a disproportionate effect on minority populations, even though 35 percent of African Americans in Atlanta do not own cars (Bullard et al., 2000). Because members of minority groups do more walking than others, they are at greater risk for pedestrian injuries in a city that eschews safe pedestrian environments. The ever-increasing sprawl of the city, with its growing congestion and continued destruction of green space, places costs on its population that are disproportionately borne by minority members.

Some panelists agreed that addressing environmental justice and forging solutions to the problems of environmental health require reaching across boundaries of race, ethnicity, culture, profession, neighborhood, and county to create a dialogue. Progress can be made when we build networks based on trust and reciprocity and work together toward common goals, many speakers suggested.

The environmental justice movement has borrowed much from Native Americans and indigenous people in terms of living in harmony with nature, explained Bullard. Several speakers concurred that for our long-term health as a species, we must view ourselves as a subset of the environment and make the health of the natural environment, and our place in it, our paramount goal. To create a bridge between the individualistic spirit in American society and the notion of social capital as valuable for the whole society, we must realize that each individual must make compromises for the common good. Accompanying this view must be the understanding that what is good for the whole society is also good for the individual. As educators, environmentalists, and health professionals, it is incumbent on us to create this vision in our communities.

References

Blair S, Kampert J, Kohl H III, Barlow C, Macera C, Paffenbarger R Jr., Gibbons L. 1996. Influences of cardiorespiratory fitness and other precursors on cardiovascular disease and all-cause mortality in men and women. *Journal of the American Medical Association* 276:205–210.

Brookings Institution. 2002a. Moving Beyond Sprawl: The Challenge for Metropolitan Atlanta. Available on-line: http://www.brook.edu/dybdocroot/urban/atlanta/population.htm [accessed June 11, 2002].

Brookings Institution. 2002b. Chapter IV. Behind the Trends: Lessons from Atlanta's History. Available on-line: http://www.brook.edu/dybdocroot/es/urban/atlanta/lessons.htm [accessed July 29, 2002].

Bullard R. 1996. *Unequal Protection: Environmental Justice and Communities of Color*. San Francisco: Sierra Club Books.

Bullard R. 2000. *Dumping in Dixie: Race, Class and Environmental Quality*, 3rd ed. Boulder, CO.

Bullard R, Johnson G, Torres A. 2000. *Sprawl City: Race, Politics and Planning in Atlanta*. Washington, DC: Island Press.

Bunker J, Frazier H, Mosteller F. 1994. Improving health: Measuring effects of medical care. *Milbank Quarterly* 72:225–258.

Calthorpe P. 1993. The Next American Metropolis: Ecology, Community, and the American Dream. Princeton: Princeton Architectural Press.

Camacho T, Roberts R, Lazarus N, Kaplan G, Cohen R. 1991. Physical activity and depression: Evidence from the Alameda County study. *American Journal of Epidemiology* 134:220–231.

Carson R. 1962. *Silent Spring*. Boston: Houghton Mifflin Company.

Centers for Disease Control and Prevention. 1999a. Ten great public health achievements: United States 1900–1999. *Morbity and Mortality Weekly Reports* 48:241–243.

Centers for Disease Control and Prevention. 1999b. Achievements in public health, 1990-1999. Motor vehicle safety: A 20th century public health achievement. *Morbidity Mortality Weekly Reports* 48:369–374.

Centers for Disease Control and Prevention. 2002a. Ten Leading Causes of Death, United States: 1999, All Races, Both Sexes. Available on-line: http://webapp.cdc.gov/cgi-bin/broker.exe [accessed June 11, 2002].

Centers for Disease Control and Prevention, National Center for Chronic Disease Prevention and Health Promotion. 2002b. Causes of Death, United States, 1900 and 1998. Available on-line: http://www.cdc.gov/nccdphp/upo/graph1.htm [accessed July 26, 2002].

REFERENCES

Centers for Disease Control and Prevention, National Center for Health Statistics. 2002c. Fast Stats A to Z: Life Expectancy. Available on-line: http://www.cdc.gov/nchs/fastats/lifexpec.htm [accessed July 26, 2002].

Centers for Disease Control and Prevention, National Center for Health Statistics, Vital Statistics System. 2002d. Available on-line: http://www.cdc.gov/nchs/nvss.htm [accessed August 1, 2002].

Cohen BA, et al. 1997. *Mean Streets. Pedestrian Safety and Reform of the Nation's Transportation Law.* Washington: Surface Transportation Policy Project and Environmental Working Group. Available on-line: http://www.ewg.org/pub/home/Reports/meanstreets/mean.html.

Committee of the Environmental and Occupational Health Assembly. American Thoracic Society. 1996. Health effects of outdoor air pollution. *American Journal Respiratory Critical Care Medicine* 153:3–50, 477–498.

Cubbin C, LeClere F, Smith GS. 2000. Socioeconomic status and the occurrence of fatal and nonfatal injury in the United States. *Am J Public Health* 90:70-77.

Dunn A, Trivedi M, O'Neal H. 2001. Physical activity dose–response effects on outcomes of depression and anxiety. *Medicine & Science in Sports & Exercise* 33:S587–S597.

Environmental Protection Agency. 2000. Assessing New York City's Watershed Protection Program: The 1997 Filtration Avoidance Determination Mid-Course Review for the Catskill/Delaware Water Supply Watershed. Available on-line: http://www.epa.gov/region02/water/nycshed/fadmidrev.pdf [accessed June 12, 2002].

Etzioni A. 1993. *The Spirit of Community: The Reinvention of American Society.* New York: Crown Publishers.

Foote SM, Etheredge L. 2000. Increasing use of new prescription drugs: A case study. *Health Affairs* 19:165-170.

Friedman M, Powell K, Hutwagner L, Graham L, Teague W. 2001. Impact of changes in transportation and commuting behaviors during the 1996 Summer Olympic Games in Atlanta on air quality and childhood asthma. *Journal of the American Medical Association* 285:897–905.

Frumkin H. 2001. Beyond toxicity: Human health and the natural environment. *American Journal of Preventive Medicine* 20(3):234-240.

Georgia Environmental Facilities Authority. 1999. *Builder's Guide to Energy Efficient Homes in Georgia.* Atlanta, GA: Southface Energy Institute Inc.

Global Hydrology and Climate Center. 2002. Heat Island. Available on-line: http://wwwghcc.msfc.nasa.gov/urban/urban_heat_island.html [accessed June 11, 2002].

Grosse S, Matte T, Schwartz J, Jackson R. 2002. Economic gains resulting from the reduction in children's exposure to lead in the United States. *Environmental Health Perspectives* 110:563–569.

Hanzlick R, et al. 1999. Pedestrian fatalities—Cobb, DeKalb, Fulton, and Gwinnett Counties, Georgia, 1994-98. *Morbid Mortal Weekly Report* 48:601-05.

Harding R., Morgan F., Indermaur D., Ferrante A, Blagg H. 1998. Road rage and the epidemiology of violence: Something old, something new. *Studies on Crime and Crime Prevention* 7:221–228.

Hartley L, el Hassani J. 1994. Stress, violations and accidents. *Applied Ergonomics* 25:221–230.

Horne C. 2001. Social capital in metropolitan Atlanta. Available on-line: http://www.atlcf.org/New%20Social%20Capital%20Report.pdf.

Jakab G, Spannhake E, Canning B, Kleeberger S, Gilmour M. 1995. The effects of ozone on immune function. *Environmental Health Perspectives* 103:77–89.

Kaplan R. 1992. The psychological benefits of nearby nature. Pp. 125-133 In: Relf D (ed.) *The Role of Horticulture in Human Well-Being and Social Development: A National Symposium.* Portland, OR: Timber Press.

Knack S, Keefer P. 1997. Does social capital have an economic payoff? A cross-country investigation. *Quarterly Journal of Economics* 112:1251-1288.

Koslowsky M, Kluger A, Reich M. 1995. *Commuting Stress: Causes, Effects, and Methods of Coping*. New York: Plenum Press.

Kuczmarski R, Flegal K, Campbell S, Johnson C. 1994. Increasing prevalence of overweight among US adults. The National Health and Nutrition Examination Surveys, 1960 to 1991. *Journal of the American Medical Association* 272:205–211.

Kunstler JH. 1996. Home from Nowhere: Remaking Our Everyday World for the Twenty-first Century. New York: Simon and Schuster.

LaPorta R, Lopez-de-Silanes F, Shleifer A, Visny R. 1997. Trust in large organizations. *American Economic Review Papers and Proceedings* 87:333–338.

Lee C. 1992. Proceedings: *The First National People of Color Environmental Leadership Summit*. New York: United Church of Christ Commission for Racial Justice (preamble).

Lee I, Paffenbarger R. 2000. Associations of light, moderate, and vigorous intensity physical activity with longevity. The Harvard Alumni Health Study. *American Journal of Epidemiology* 151:293–299.

Leikauf G, Simpson L, Santrock J, Zhao Q, Abbinante-Nissen J, Zhou S, Driscoll K. 1995. Airway epithelial cell responses to ozone injury. *Environmental Health Perspectives* 103:9–95.

Litonjua AA, et al. 1999. Race, socioeconomic factors, and area of residence are associated with asthma prevalence. *Pediatric Pulmonology* 28(6):394-401.

Lomax T, Turner S, Margiotta R. 2001. Monitoring Urban Roadways in 2000: Using Archived Operations Data for Reliability and Mobility Measurement. Available on-line: http://tti.tamu.edu/ [accessed June 11, 2002].

Lopez J, Chalmers D, Little K, Watson S. 1998. Regulation of serotonin(1A), glucocorticoid, and mineralocorticoid receptor in rats and human hippocampus: Implications for the neurobiology of depression. *Biological Psychiatry* 43:547–573.

Lovejoy T. 2001. Biodiversity and health. In: Hanna K, Coussens, C. (eds.) *Rebuilding the Unity of Health and the Environment: A New Vision of Environmental Health for the 21st Century*. Washington, DC: National Academy Press.

Marosi R. 1999. Pedestrian deaths reveal O.C.'s car culture clash; Safety: Latinos, 28% of Orange County's population, are victims in 40% of walking injuries, 43% of deaths. *Los Angeles Times*, November 28, p. 1.

Martinez A. 1990. Injury control: A primer for physicians. *Annals of Emergency Medicine*. 19:72–77.

McCann B, DeLille B. 2000. *Mean Streets 2000. Pedestrian Safety, Health and Federal Transportation Spending*. Washington: Surface Transportation Policy Project.

Meriwether J, Mitigate M (eds.). 1988. In: *Lion in the Garden: Interviews with William Faulkner, 1926–1962*. New York: Random House.

Metzger R, et al. 1995. Environmental health and Hispanic children. *Environ Health Persp* 103 Suppl 6:539-50.

Mizell L. 1997. Aggressive driving. In: Aggressive Driving: Three Studies. AAA Foundation for Traffic Safety. Available on-line: http://www.aaafoundation.org/resources/index.cfm?button=agdrtext [accessed June 11, 2002].

Mo R, Wilkie C. 1997. Changing Places: Rebuilding Community in the Age of Sprawl. New York: Henry Holt and Co.

Mokdad AH, Serdula MK, Dietz WH, Bowman BA, Marks JS, Koplan JP. 1999. The spread of the obesity epidemic in the United States, 1991–1998. *Journal of the American Medical Assocation* 282:1519–1522.

Mokdad AH, Ford ES, Bowman BA, Nelson DE, Engelgau MM, Vinicor F, Marks JS. 2000. Diabetes trends in the U.S.: 1990–1998. *Diabetes Care* 23:1278–1283.

Moore EO. 1981. A prison environment's effect on healthcare service demands. *Journal of Environmental Systems* 11:17–34.

Moreno S, Sipress A. 1999. Fatalities higher for Latino pedestrians; Area's Hispanic immigrants apt to walk but unaccustomed to urban traffic. *Washington Post*, August 27. p. A01.

REFERENCES

Must A, Spadano J, Coakley E, Field A, Colditz G, Dietz W. 1999. The disease burden associated with overweight and obesity. *Journal of the American Medical Association* 282:1523–1529.

Nadel E, Cullen MR. 1994. Thermal stressors. In: Rosenstock L, Cullen MR. Textbook of *Clinical Occupational and Environmental Medicine*. Philadelphia: Saunders, pp. 658-666.

National Heart, Lung, and Blood Institute Working Group. 1995. Respiratory diseases disproportionately affect minorities. *Chest* 108:1380-1392.

National Highway Traffic Safety Administration. 1999. *Traffic Safety Facts 1998. A Compilation of Motor Vehicle Crash Data from the Fatality Analysis Reporting System and the General Estimates System.* DOT HS 808 983. Washington, DC.

National Highway Traffic Safety Administration. 2000. *Traffic Safety Facts 1999. A Compilation of Motor Vehicle Crash Data from the Fatality Analysis Reporting System and the General Estimates System.* DOT HS 809 100. Washington, DC.

National Institutes of Health. 1998. *Clinical Guidelines on the Identification, Evaluation, and Treatment of Overweight and Obesity in Adults.* Bethesda, MD: Department of Health and Human Services; National Heart, Lung, and Blood Institute.

National Institutes of Health Consensus Development Panel on Physical Activity and Cardiovascular Health . 1996. NIH Consensus Conference: Physical activity and cardiovascular health. *JAMA* 276:241-246.

Odum EP. 1998. *Ecological Vignettes: Ecological Approaches to Dealing with Human Predicaments*. New York: Harwood Academic Publishers.

Olfson M, Marcus S, Pincus H, Zito J, Thompson J, Zarin D. 1998. Antidepressant prescribing practices of outpatient psychiatrists. *Archives of General Psychiatry* 55:310–316.

Ortega y Gasset. 2000. *Meditations on Quixote*. Urbana, IL: University of Illinois Press.

Pate RR, et al. 1995. Physical activity and public health: a recommendation from the Centers for Disease Control and Prevention and the American College of Sports Medicine. *JAMA* 273:402-407.

Persky VW, et al. 1998. Relationships of race and socioeconomic status with prevalence, severity, and symptoms of asthma in Chicago school children. *Ann Allergy Asthma Immunol* 81(3):266-71.

Putnam R. 2000. *Bowling Alone: The Collapse and Revival of American Community*. New York: Simon & Schuster.

Quattrochi D. 2000. Here Comes Urban Heat. In the article, Dr. Quattrochi cites an analysis of photographs and data collected by the National Aeronautics and Space Administration's Landsat 7 from 1973 to 1992. Available on-line: http://science.nasa.gov/headlines/y2000/essd16mar%5F1m.htm [accessed July 29, 2002].

Quattrochi D, Luvall J. 1999. *Atlanta Land-Use Analysis: Temperature and Air-Quality Project.* NASA Global Hydrology and Climate Center, Marshall Space Flight Center, Huntsville, AL.

Quattrochi DA, Luvall JC, Rickman DL, et al. 1998. Project Atlanta (Atlanta land use analysis: temperature and air quality) [microform]: A study of how the urban landscape affects meteorology and air quality through time. Huntsville, AL.: National Aeronautics and Space Administration Global Hydrology and Climate Center.

Rathbone D, Huckabee J. 1999. Controlling Road Rage: A Literature Review and Pilot Study. AAA Foundation for Traffic Safety. Availabe on-line: http://www.aaafoundation.org/pdf/RoadRageFinal.PDF [accessed June 11, 2002].

Saguaro Seminar. 2002. Civic Engagement in America. Available on-line: http://www.bettertogether.org/pdfs/bt_1_29.pdf [accessed August 1, 2002].

Sesso H, Paffenbarger R, Lee I. 1998. Physical activity and breast cancer risk in the College Alumni Health Study (United States). *Cancer Causes & Control* 9:433–439.

Sesso HD, et al. 1999. Physical activity and cardiovascular disease risk in middle-aged and older women. *Am J Epidemiol* 150(4):408-16.

Shaper A, Wannamethee S, Walker M. 1997. Body weight: Implications for the prevention of coronary heart disease, stroke, and diabetes mellitus in a cohort study of middle aged men. *British Medical Journal* 314:1311–1317.

Ulrich RS. 1984. View through a window may influence recovery from surgery. *Science* 224:420–421.

United Kingdom National Health Service. 2000. Exercise Therapy. Available on-line: http://cebmh.warne.ox.ac.uk/cebmh/elmh/depression/treatment/exercise1.html [accessed June 11, 2000].

U.S. Census Bureau. 2000a. Annual Projections of the Total Resident Population as of July 1: Middle, Lowest, Highest and Zero International Migration Series, 1999 to 2100. Available on-line: http://www.census.gov/population/www/projections/natsum.html [accessed June 12, 2002].

U.S. Census Bureau. 2000b. Population trends in metropolitan areas and central cities, 1990–1998. Available on-line: http://www.census.gov/prod/2000pubs/p25-1133.pdf [accessed July 29, 2002].

U.S. Census Bureau, Population Division, Population Projections Branch. 2000c. National Population Projections. I. Summary Tables. Available on-line: http://www.census.gov/population/www/projections/natproj.html [accessed July 29, 2002].

U.S. Department of Energy. 2002. Energy Savers: Tips on Saving Energy and Money at Home. Available on-line: http://www.eren.doe.gov/consumerinfo/energy_savers/summer/summer.html [accessed July 29, 2002].

U.S. Department of Health and Human Services 1996. *Physical Activity and Health: A Report of the Surgeon General.* Washington, DC.

Wannamethee S, Shaper A, Walker M, Ebrahim S. 1998. Lifestyle and 15-year survival free of heart attack, stroke, and diabetes in middle-aged British men. *Archives of Internal Medicine* 158:2433–2440.

Wannamethee SG et al. 1999. Physical activity and the prevention of stroke. *J Cardiovasc Risk* 6(4):213-6.

Wei M., Kampert J, Barlow C, Nichaman M, Gibbons L,. Paffenbarger R, Blair S. 1999. Relationship between low cardiorespiratory fitness and mortality in normal-weight, overweight, and obese men. *Journal of the American Medical Association* 282:1547–1553.

Wernett, D, Nieves, L. 1992. Breathing polluted air: Minorities are disproportionately exposed. *EPA Journal* 18:16–17.

Willett W, Dietz W, Colditz G. 1999. Guidelines for healthy weight. *New England Journal of Medicine* 341:427–434.

World Health Organization. 1986. Constitution. In: *World Health Organization: Basic Documents.* Geneva, Switzerland.

Appendix A

Agenda

REBUILDING THE UNITY OF HEALTH AND THE ENVIRONMENT IN THE SOUTHEASTERN UNITED STATES

Cosponsored by:
Institute of Medicine
Rollins School of Public Health, Emory University
National Center for Environmental Health, Centers for Disease Control and Prevention
Georgia Institute of Technology
National Institute of Environmental Health Sciences, National Institutes of Health

June 27, 2001
Cyprus Room
The Carter Center
Atlanta, GA

9:00 a.m. **Welcome**

9:05 a.m. **Opening**
Dr. Joseph Lowery
Chairman
Georgia's Coalition for the People's Agenda

9:20 a.m. **Rebuilding the Unity of Health and the Environment**
Richard Jackson, M.D., M.P.H
Director, National Center for Environmental Health
Centers for Disease Control and Prevention

9:50 a.m. **Panel I: The Natural Environment**

Moderator: John Sibley
President, The Georgia Conservancy

Robert Kerr
Director, Pollution Prevention Assistance Division
Georgia Department of Natural Resources

Eugene Odum, Ph.D.
Callaway Professor Emeritus, and Director Emeritus of Institute
  of Ecology, University of Georgia

Wayne Alexander, M.D.
R. Bruce Logue Professor and Chairman of Medicine
Emory University

Respondents:
Sally Bethea
Director
Upper Chattahochee Riverkeeper

Samuel H. Wilson, M.D.
Deputy Director
National Institute of Environmental Health Sciences
National Institutes of Health

10:40 a.m. **Audience Participation**

11:15 a.m. **Break**

11:30 a.m. **Panel II: The Built Environment
(Buildings, Roads, Transport, and Energy)**

Moderator: Lynn Goldman, M.D.
Professor, Johns Hopkins University, School of Public Health

Ricardo Martinez, M.D.
Chief Executive Officer and President
Safety Intelligence Systems

Michael Kilgallon
Owner
The Pacific Group

Charles Liotta, Ph.D.
Vice Provost for Research, Dean of Graduate School
Georgia Institute of Technology

Dennis Creech
Executive Director
Southface Energy Institute

Charles Goodman, Ph.D.
Senior Vice President for Research and Environmental Affairs
Southern Company

Respondents:
Linda DiSantis, J.D.
Corporate Compliance Manager
United Parcel Service

Lawrence Frank, Ph.D.
Associate Professor
Georgia Tech

12:30 p.m. **Audience Discussion**

1:00 p.m. **Lunch (provided for all participants)**

12:00 p.m. **Afternoon Welcome**
The Honorable Paul G. Rogers, J.D.
Chair
Roundtable on Environmental Health Sciences, Research, and Medicine

Jeffrey P. Koplan, M.D., M.P.H.
Director
Centers for Disease Control and Prevention

2:20 p.m. **Panel III: Social Environment: From Social Justice to Social Capital**

Moderator: Baruch Fischhoff, Ph.D.
Professor of Social and Decision Sciences
Carnegie Mellon University

Elliott Sclar, Ph.D.
Professor of Urban Planning and Public Policy
Columbia University

Winsome Hawkins
Vice-President of Programs and Initiatives
The Community Foundation for Greater Atlanta, Inc.

Robert Bullard, Ph.D.
Director of the Environmental Justice Resource Center
Clark Atlanta University

3:10 p.m.   **Audience Participation**

3:30 p.m.   **Summation**

Howard Frumkin, M.D., M.P.H., Dr. P.H.
Professor and Chair
Department of Occupational and Environmental Medicine
Rollins School of Public Health
Emory University

4:00 p.m.   **Adjourn**

Appendix B

Speakers and Panelists

Wayne Alexander, M.D., Ph.D.
R. Bruce Logue Professor and
 Chairman
Emory University

Robert Bullard, Ph.D.
Director of the Environmental Justice
 Resource Center
Clark Atlanta University

Dennis Creech
Executive Director
Southface Energy Institute

Linda DiSantis, J.D.
Corporate Compliance Manager
United Parcel Service, Inc.

**Lawrence D. Frank, Ph.D., RLA,
 AICP**
Assistant Professor
Georgia Institute of Technology

**Howard Frumkin, M.D., M.P.H.,
 Dr.P.H.**
Professor and Chair
Department of Environmental and
 Occupational Health, Emory
 University Rollins School of
 Public Health

Charles Goodman, Ph.D.
Senior Vice President for Research
 and Environmental Affairs
Southern Company

Lynn Goldman, M.D.
Professor
Johns Hopkins University,
 Bloomberg School of Public
 Health

Winsome Hawkins
Vice President
The Community Foundation for
 Greater Atlanta

Richard Jackson, M.D., M.P.H.
Director
National Center for Environmental
 Health, Centers for Disease
 Control and Prevention

Matthew Kales
Upper Chattahochee Riverkeeper

Robert Kerr
Director
Georgia Department of Natural
 Resources

Michael Kilgallon, MBA
Chair
Government Affairs Council of the Greater Atlanta Homebuilders Association

Jeffrey Koplan, M.D., M.P.H.
Director
Centers for Disease Control and Prevention

Charles Liotta, Ph.D.
Vice Provost for Research, Dean of Graduate Studies
Georgia Tech

Joseph Lowery, BA, BD, DD
President
Georgia Coalition for the People's Agenda

Ricardo Martinez, M.D.
Chief Executive Officer and President
Safety Intelligence Systems

Eugene Odum, Ph.D.
Alumni Foundation Distinguished Professor Emeritus, Callaway Professor Emeritus, and Director Emeritus
Institute of Ecology

The Honorable Paul Rogers, J.D.
Partner
Hogan and Hartson

Elliott Sclar, Ph.D.
Professor of Urban Planning and Public Policy
Columbia University

John Sibley III, J.D.
President
The Georgia Conservancy

Samuel H. Wilson, M.D.
Deputy Director
National Institute of Environmental Health Sciences
Deputy Director
National Toxicology Program, National Institutes of Health

Appendix C

Meeting Participants

Edwin Akins
Smith Dalia Architects

Heather Alhadeff
Federal Highway Administration

Lee Anthony

Ed Arnold
Physicians for Social Responsibility

Gerry Barrett

Karsten Baumann
Georgia Tech

Terrilyn Bayne
League of Conservation Voters
 Education Fund

Carolyn Beeker
Centers for Disease Control and
 Prevention

Ben Bellows
Centers for Disease Control and
 Prevention

Amy Berlin
Centers for Disease Control and
 Prevention

Kay Beynart
North Buckhead Civic Association

Achal Bhatt

Sue Binder
Centers for Disease Control and
 Prevention

Rob Blake
Dekalb County Board of Health

Tom Bluewolf
Creek Nation/Earthkeepers

Marnie Boardman
Centers for Disease Control and
 Prevention

Nathan Boddie
Trees Columbus, Inc.

Gail Bosley
Public Health Practice Program
 Office

Caroline Bragdon
Centers for Disease Control and
 Prevention

Spencer Brewer
Piedmont Medical Center/Emory University, School of Medicine

C. Brock
Centers for Disease Control and Prevention

Julia Bryson
Agency for Toxic Substances and Disease Registry

Emily Burnett
University of the South/Sewanee, TN

Nancy Burnett
Centers for Disease Control and Prevention

Heather Carter
Centers for Disease Control and Prevention

William Carter
Agency for Toxic Substances and Disease Registry

Michael Chang
Georgia Tech

Keera Cleare
U.S. Army Environmental Policy Institute

Mindy Clyne
Office of Genetics and Disease Prevention, Centers for Disease Control and Prevention

Gini Cogswell
Univeristy of Georgia, Institute of Ecology

Daniel Cohan
Georgia Tech

Gregory Cooper
Earth Resource Group

Helen Crawford
Lord Aeck Sargent Architects

Susan Cummins
National Center for Environmental Health, Centers for Disease Control and Prevention

Andrew Dannenberg
Centers for Disease Control and Prevention

Allen Dearry
National Institute of Environmental Health Sciences/National Institutes of Health

Robert Delaney
Centers for Disease Control and Prevention

Francis Desiderio
Office of University–Community Partnerships, Emory University

Bob Donaghue
Georgia Pollution Prevention Assistance Division

Paul Dorsey

Richard Ehrenberg
National Center for Injury Prevention and Control, Centers for Disease Control and Prevention

Cary M. Ellis
Winter Environmental Services, Inc.

Dolly Evans
Small Business Services

MEETING PARTICIPANTS

Tom Ferguson
Physicians for Social Responsibility

Julie Fishman
Centers for Disease Control and Prevention

Gray Folger

Suzanne Folger
Centers for Disease Control and Prevention

Sara Fountain

Erica Frank
Physicians for Social Responsibility

Jonlyn Freeman
Village Habitat Design

Amy Funk
Rollins School of Public Health, Emory University

Debra Gable
Agency for Toxic Substances and Disease Registry

Paul Garbe
Centers for Disease Control and Prevention

Kenneth Garrard
Earth Resource Group

Beth Gavrilles
University of Georgia Institute of Ecology

Trey Gibbs
Georgia Environmental Organization

Siobhan Gilchrist
Centers for Disease Control and Prevention

Donna Gillroy
Gillroy & Associates

Amanda Gonzalez
Agency for Toxic Substances and Disease Registry

Marlin Gottschalk
Georgia Department of Natural Resources

Molly Greaney
Centers for Disease Control and Prevention

Daniel Green
Centers for Disease Control and Prevention

Heather Gregory
Centers for Disease Control and Prevention

Jennifer Grubbs
Agency for Toxic Substances and Disease Registry

Jewell Grubbs
U.S. Environmental Protection Agency

George Hadley
University of Georgia

Robert Hahn
Centers for Disease Control and Prevention

Rebecca Hart
Centers for Disease Control and Prevention

Holley Henderson
TVS Architects

Jason Henderson
University of Georgia

Erin Hennessy

Wayne Henry
Centers for Disease Control and Prevention

Rosemarie Henson
National Center for Environmental Health

Jerry Hershovitz
Centers for Disease Control and Prevention

Azania R. Heyward-James
Centers for Disease Control and Prevention

Nancy Hood
Rollins School of Public Health

Steve Hortin
U.S. Department of Energy

Elizabeth Howze
Centers for Disease Control and Prevention

Bryan Hudson
American Friends Service Committee

Elayne Hunter
The Outdoor Activity Center

Gladys Ibanez
Mexican American Legal Defense and Educational Fund

Carl James
Agency for Toxic Substances and Disease Registry

Pilar Jan
Penn Southern Environmental Law Center

Robert Jarrett
U.S. Army Environmental Policy Institute

Mark Jensen
Cooper Carry

Henry Kahn
Centers for Disease Control and Prevention

Maisha Kambon
Centers for Disease Control and Prevention

Dafna Kanny
Agency for Toxic Substances and Disease Registry

Mark Kashdan
Centers for Disease Control and Prevention

Susan Katz
Centers for Disease Control and Prevention

Carla Keplinger
U.S. Department of Veterans Affairs

Rosemary Kernahan
Atlanta Department of Community Affairs

Nikki Kilpatrick
Centers for Disease Control and Prevention

MEETING PARTICIPANTS

Rita Kilpatrick
Georgians for Clean Energy

Joan O. King

Jeffrey Kirsch

Lisa Kruse
University of Georgia

Adele Kushner
Action for a Clean Environment

Traci Leath
U.S. Department of Energy

Sarah Levin
Centers for Disease Control and Prevention

Michael Lewyn
John Marshall Law School

Susan Lockhart
Centers for Disease Control and Prevention

Jada Locklear
National Center for HIV, STD, and TB Prevention, Centers for Disase Control and Prevention

Judy Long
Hill Street Press

Jessica Lowy
Centers for Disease Control and Prevention

James Mack
State of Oregon

Ferhan Manas
Peachtree Psychoeducational Service

John Mann
Agency for Toxic Substances and Disease Registry

Randall Manning
Georgia Environmental Protection Division

Marilyn B. Marks
Earth Resource Group

Joseph Martin
National Park Service

Karen Maschke

Kaly McKibben
Atlanta Girls' School

Linda McKibben
Agency for Toxic Substances and Disease Registry

Patrick Meehan
Centers for Disease Control and Prevention

Debbie Moll
Centers for Disease Control and Prevention

Lawrence Morris
University of Georgia

Belinda Morrow
2M Design Consultants, Inc.

Leonard Morrow
2M Design Consultants, Inc.

Anthony D. Moulton
Centers for Disease Control and Prevention

Mary Ellen Myers
Action for a Clean Environment

Melvin L. Myers
Emory University

Vincent Nathan
Agency for Toxic Substances and Disease Registry

Andrea Neiman
Centers for Disease Control and Prevention

Jennifer Neiner
Centers for Disease Control and Prevention

Tom Nessmsith
U.S. Environmental Protection Agency

Karen Nozik
Rails to Trails Conservancy

Ralph O'Connor
Agency for Toxic Substances and Disease Registry

Donna Orti
Agency for Toxic Substances and Disease Registry

Yamil Padilla

Daniel Parshley
Glynn Environmental Coalition

James Patterson
Georgia Environmental Organization

Seleda M. Perryman
Centers for Disease Control and Prevention

Phillip H. Pfeifer
Gwinnett County Department of Public Utilities

Anne Pollock
Centers for Disease Control and Prevention

Kristin Pope
Centers for Disease Control and Prevention

Cynthia Poselenzny
Packer Industries, Inc.

Judith Qualters
Centers for Disease Control and Prevention

Matthew Radune
Village Habitat Design

Barbara Reeves
Chattahoochee Nature Center

Steven Reynolds
Centers for Disease Control and Prevention

Betsy Rivard
Women's Action for New Directions

Jennifer Robinson
Centers for Disease Control and Prevention

Felix Rogers
Centers for Disease Control and Prevention

Lee Ross
Earth Resource Group

MEETING PARTICIPANTS

Raquel Sabogal
Centers for Disease Control and Prevention

Polly Sattler
The Georgia Conservancy

Thomas Sayre
Sizemore Floyd

Amanda Schofield
Centers for Disease Control and Prevention

Jessica Shisler
Centers for Disease Control and Prevention

Thomas Sinks
Centers for Disease Control and Prevention

Jeffrey Sitterle
Georgia Tech Research Institute

Phil Sparling
Georgia Tech

Susie Spivey-Tilson

Jan Stansell
Centers for Disease Control and Prevention/National Center for Chronic Disease Prevention and Health Promotion

Sarah Statham

Pat Stevens
Atlanta Regional Commission

Brenda Stokes
Beers Construction Co.

Rebecca Stoner
Smith Dalia Architect

Sylvia Struck

Daniel Swartz
Children's Environmental Health Network

Mildred Lee Tanner
South Carolina Department of Health and Environmental Control

Gerald Teague
Emory University

Claude E. Terry
CTA Environmental, Inc.

Linda Thomas
U.S. Environmental Protection Agency

Pamella Thomas

Francisco A. Tomei
Torres Office of Urban Affairs

Connie Tucker
Southern Organizing Committee

Pamela Tucker
Agency for Toxic Substances and Disease Registry

James Tullos
Agency for Toxic Substances and Disease Registry

Dan Turner
Daniel Turner Builders, Inc.

Janet Valente
University of Georgia

Susan Varlamoff
University of Georgia

Theodosia Wade
Oxford College of Emory University

Justin Waltz
Centers for Disease Control and Prevention

Andrew Watkins
Centers for Disease Control and Prevention

Tom Weyandt
Atlanta Regional Commission

Scott Wheeler
Cooper Carry

Randall White

Erin Wieckert

Marcus Wilner
Federal Highway Administration

Anne Wilson
Centers for Disease Control and Prevention

Dorothea Wilson

Geraldine Wolfle

David Word
Georgia Environmental Protection Divisin

Russell Wright, Jr.
U.S. Environmental Protection Agency, Region 4

Lynn Zanardi
Centers for Disease Control and Prevention

Max Zarate

Corey Zetts

Amy Zlot
Centers for Disease Control and Prevention